Dr Una Coales's MRCGP CSA Book

Second Edition

Una Coales BA (Hon) MD FRCSEd FRCSEd (ENT)
FRCGP DRCOG DFFP PGCertMedEd

www.lulu.com

First published in Great Britain by www.lulu.com 2010.
Second edition published by 2013.

Published and distributed by Lulu.com
3101 Hillsborough Street,
Raleigh, NC 27607, USA.
Website: www.lulu.com

Contents

Preface

The n (ew) MRCGP exam, comprising of AKT (applied knowledge test) and CSA (clinical skills assessment) components, commenced October 2007 and represents the licensing exam for all UK GP specialist trainees. Passing the MRCGP is required to obtain your certificate of completion of GP specialist training (CCST) in order to work in the NHS as a fully qualified GP. Trainees who fail either component of this exam may find themselves appearing before Deanery review boards to plead their case to receive training extensions. Deaneries argued passing the AKT module was down to book-learning. However, Deaneries decided to grant automatic 6-month training extensions as GP trainees started failing the CSA (clinical skills assessment) module. The CSA module alone costs £1,525 (2013). Record 5th-timers re-sitting AKT and 5th-timers (5th granted on humanitarian grounds) re-sitting CSA are facing substantial financial and deanery pressures to pass nMRCGP. Is this a case that not all GP trainees are suited to complete training to become GPs or is there something else that could account for these repeated failures?

This book is the culmination of many years of teaching over 4,000 'star' GP trainees amidst 'problem learners.' In my opinion, they are both geniuses; the differences lie in attitudes, confidence levels, baseline knowledge, dedication and determination. All this can be changed with a trainer willing to take on Harden's 12 roles of a teacher, to understand Honey and Mumford's different learning styles to accommodate each and all and to transform learners into experts with kindness, patience and a syllabus which offers both Entwistle's strategic and deep knowledge.

In 1993, as a US surgeon, I sat my first UK postgraduate exam, the PLAB exam with its 5-modules, fail one module, re-sit all policy. I made a promise to write a book on this exam if I ever passed! As I progressed through most of the UK postgraduate exams (including the masochistic negatively-marked FRCS exams), I wrote exam book aides along the way to help colleagues. In 2002, as I sat the MRCGP exam at the Royal Horticultural Hall amidst 400 GPs, I wondered what exactly had I revised over 6 weeks as nothing seemed relevant to the actual MCQ paper and who was manning our empty GP surgeries? So I wrote books for the Royal Society of Medicine on the MCQ, written and oral and video modules of the MRCGP exam based on my revision notes.

There were no RCGP courses or books on the Simulated Surgery module and not surprisingly, the RCGP published historical pass rates between 56% and 60%. These GPs (both UK and International who could not video-tape) needed help and so in 2004, I began teaching all 5 modules of the old MRCGP (MCQ, written, oral, video and simulated surgery) and now solely on the new MRCGP exam (AKT and CSA), as the old exam phased out.

The AKT is modelled on the old MCQ paper but is computerised (200 MCQs in 3 hours) and the CSA is modelled on the old Simulated Surgery module (with now 13 marked stations of 10-minute consultations as of September 2010). The RCGP published Spring 2008 AKT results of 1st time pass rate 85.5% and 2nd time 66.3%, and the CSA results of 1st time pass rate 81.3% and 2nd time pass rate 63.7%. Interestingly, the RCGP also published the White pass rate for CSA of 93.3% and Asian pass rate of 63.9%.

From September 2010, the CSA marking system changed to borderline group method. The little 'n' was dropped so it is now again called the MRCGP exam. In essence at the end of each exam day, the examiners will meet and decide how many total points equate to an overall pass. This pass mark will differ each day depending on the strength of the cohort sitting and varies between 73 and 77/117 marks with each station being marked out of a score of 9 (6 being a pass). As this is now a UK GP licensing exam, it means more than ever, that one has to perfect one's consultation skills as a practitioner. There can be no room for error or nerves. You will not know by the end of the day how you faired!

4

This 2nd edition updates the original August 2010 book and offers both theories for and solutions on how to combat a concept called 'unconscious bias' in a subjective human-marked CSA exam with no shadow examiner and no video monitoring for CSA licensing exam appeals. The CSA pass rates for black and minority ethinic UK and IMG (international medical graduates) have shown a stark difference compared with white UK pass rates, yet all have undergone the same rigorous competitive UK entrance exams into GP specialty training programmes and all have completed 3-3.5 years of GP speciality training. The CSA exam pass rate for 2010-2011 published by the Royal College of General Practitioners was 96.1% for white UK graduates (dropping to 84.7% for Asians, 77.3% for Chinese and 66.7% for Black UK graduates) and as low as 36.9% for Asian and 35.1% for Black International Medical Graduates (IMGs), yet all have sat CSA during their final year of UK GP specialty training and many have had no concerns flagged by their GP trainers or their NHS patients. 612 ethnic UK and IMG GP trainees failed their first attempt at CSA in 2011-12.

Be enlightened, transform your GP consulting style and pass CSA! This book is an adjunct to my CSA courses which may be booked on www.mrcgpcourses.co.uk.

The book is dedicated to my baby brother Henry who died of a missed NHS GP diagnosis of intussusception and to Dr John Coales, winner of the Guys Hospital Gold Medal for the First Prize in the practice of medicine in 1837 (great great great uncle to my daughters). Sadly he died of tuberculosis shortly after taking up his post as a surgeon in Harley Street.

Dr Una Coales MD FRCSEd FRCSEd (ENT) FRCGP DRCOG DFFP PGCertMedEd
www.mrcgpcourses.co.uk
mrcgpcourses@aol.com

May 2013

CSA Lesson One: English Manners

'I have never seen such an appalling lack of manners in all my life,' a female examiner might think subconsciously when seeing a male GP not aid an elderly lady, hobbling with a cane, to her chair, much less help her out of it. *'What a shame the culture of medicine does not include a lesson in manners?'*

Some GPs fail to introduce themselves to patients or introduce themselves as an object, *'It's Dr X here'*? Gone are the days, in which students would rise and chorus, *'**Good morning Miss Smith**,'* to their teacher, pause and wait for permission to sit down. Try *'Good morning Miss Smith, how do you do? I'm Dr X. Would you like to come in and have a seat'*? I often tell GP trainees that it is NOT *'Hi'*, we are NOT in *'**Hi**gh school'*! And nor do we issue commands, *'Take your seat!'* It's 'have a seat' not 'take your seat'!

Please remember your 'please's' and 'thank you's'. It is polite to ask permission (use the 'modal') when addressing our esteemed patients, *'**may I** examine you please'*? How many GPs command their patients like children and for that matter, marital partners? *'Come in!'* *'The GP can't be bothered to get out of his chair and open the door for me, much less greet me with a handshake'*, the actor patient may be subconsciously thinking. *'You need to take this medicine,'* as I am your arrogant, authoritative doctor-centred GP, is the subliminal message you are conveying in CSA. A GP touches patients. Shake the hands of your actor patients.

A gentleman follows a lady to her chair, pulls it out and pushes it back in as she sits down. Female GPs we may not sit down until our patient sits down, so stand and pause. *'Would you like to sit down?'* and if the patient chooses to stand, then ask, *'May I sit down?'* and sit like a lady, legs at a diagonal. Never enter into a conversation with an angry patient with both parties standing! GPs also ask permission to remove our jackets if the room is too hot. *'Do you mind if I removed my jacket?'*

CSA Lesson Two: Too Doctor-Centred

The number one reason for failing CSA is being too doctor-centred. What does this really mean? The analogy I often use to demonstrate this is the 'waiter in a restaurant' scenario. Have you ever heard a waiter approach your table and declare, 'I think you should start with the Caesar salad. Then you will have salmon and potatoes. And for pudding, we will order you the crème brulee. Is that okay?' This usually gets peals of laughter from the audience and yet, this is how hospital doctors speak. Patients would not dare challenge a surgeon regarding a recipe to take out an appendix. However being a GP is more about shared management or empowering the patient so that he can understand his illness and is competent to be involved in decision-making regarding the management of his condition.

So how do GPs become more patient-centred?

- Never use the egotistically word 'I' or the royal 'we'. Have a go. Try to speak for 10 minutes without referring to yourself and your ego. When doctors use 'I' repeatedly, it gives the unconscious impression of arrogance.

- Neutralise your speech. Use the gender-neutral pronoun 'it.' 'It may be a good idea to...'

- Do not talk more than your patient. Running monologues are not permitted. If you talk and talk and talk, then YOU are the centre of your universe. Professor Steve Field, the Chairman of the RCGP (2007-2010) and former deputy video examiner advises that *'the patient should be the centre of your universe.'*

- Share in the dialogue. You speak for 1-2 sentences, and then the patient speaks for 1-2 sentences. This gives the impression of a shared 50:50 doctor-patient interaction.

- Customer service with a friendly smile. Yes, smile and do not look stern at patients. A stern look may be misconstrued as cold-heartedness. A genuine smile is warm, welcoming and endearing, especially when being 'assessed' by 26 strangers! The 'Mona Lisa' smile has a positive unconscious bias! Reflect the 'Mona Lisa' smile while listening silently to the actors.

- Treat patients with respect and courtesy, as you would have them treat you.

- Educate the patient without jargon so that patients may better understand their conditions and be empowered to share in decision-making. Apply the 8-year-old test. Can an 8-year-old understand you?

- Treat patients like adults and not children. Do not dictate to your patients. Offer your patients options/ choices. 'The options are 1, 2, 3. What would you like?' Good, you are now sounding like a waiter, patient-customer-centred.

- Do not lecture to your patients. After assessing motivation, if they are not ready to quit smoking, respect their choice and offer help when they are ready. Do not nag them to death!

- Avoid jargon and grandiosity. Not all patients have a university degree. Keep language simple and you may draw a basic diagram to illustrate better. Patient leaflets on www.patient.co.uk give examples of how to explain various conditions in layman's terms. This is NOT MRCP or MRCS so stop showing off with verbal diarrhoea!

- There are 2 negative phrases one must never use with a patient or they will explode with fury; '*I can't*' and '*but*'. Think customer service and how upset we get when we hear these words. So try to think of another phrase to use when letting a patient down gently or put a positive spin to your phrases. Exchange '*but*' with '*however*' if you must.

- Do not argue with your patients! This warrants a clear fail. If it has reached this dysfunctional point, stop, and take a step back. Apologise and ask again 'what were you hoping I could do for you?' There are 650 CSA cases, so remember my Golden Rule of CSA! **'Give them anything they want except for an unlicensed drug, a new prescription for benzodiazepine (may give on a weekly rx if a repeat), inappropriate sick notes (ie holiday), methadone (instead refer to local drug services) or if it is wasting NHS resources.'**

  Examples of doctor-centred speech and corrected patient-centred speech.

- '*I am going to refer you to see a hospital specialist*'. Rather ask '*may I or would you like me to refer you to the hospital specialist?*'
- '*You will start this tablet.*' Try '*Would you be happy to start a tablet to lower your blood pressure?*'
- '*I will see you next week*'. Rather ask '*may I see you next week?*' as they may be going on holiday next week. It is arrogant to presume they are free to see you at your will.

CSA Lesson Three: How to Combat Unconscious Bias

How to combat unconscious and subjective bias to succeed in CSA and for that matter a GP partnership interview? Unlike other College examinations, the CSA does not have a shadow examiner to reduce unconscious bias; marking at each of the 13 stations is done by one examiner for each case with no videos for appeals. He or she may be influenced by other GPs on the same floor role-playing the same case. Unconscious bias is a concept pioneered in 2007 by Professor Banaji in the US and the Harvard Implicit Association test is well established to test individuals for 'unconscious bias'. The CSA does not test actors or examiners.

Dress for success. The politically correct response is that it should not matter what you choose to wear, as you are marked on your clinical ability and not your dress sense. The truth is that humans form instant subjective opinions within seconds of meeting a stranger. Do you sit up straighter for a professionally-dressed patient? Dress in a suit for a week and see if your patients and colleagues treat you any differently. It is an eye-opening experience. In the old MRCGP exam, one might be asked to define professionalism. After all, we doctors hope to be regarded in the same league as barristers and bankers? Would you expect to consult with a barrister who is dressed in casual weekend attire, especially if your life is at stake?

Still not convinced? Read 'Blink' by Malcolm Gladwell or 'Thinking Fast and Slow' by Nobel Prize winner Daniel Kahneman. In the blink of an eye, 13 stranger GPs and 13 actor patients will decide whether they like you and 13 GPs decide whether you are a clear pass, pass or fail candidate.

Smile and pass. Frown and fail. In 1955, Professor Albert Mehrabian, a renowned US educational psychologist, published research that communication was 55% nonverbal, 38% paralinguistics and 7% content. What we do with our face, hands and body posture matter. In 1872, Charles Darwin published the book 'The Expression of the Emotions in Man and Animals', discussing the significance behind facial expressions and communication. What should our facial expressions reflect? Joy at being a GP, so smile, with the exception of breaking bad news.

Paralinguistics? That means tone, pitch, rate, accent, etc. Start by tape-recording your voice. A study in the States showed that surgeons with harsh tones got sued the most and that there was no correlation between being sued and competency. So much for patient satisfaction surveys! So we ensure our tone is soft (as though speaking to a child), our pitch is deep (not shrill like a nagging wife), our rate is measured and slow (not pressured), and our accent is neutral (must sound educated at the very least!). The actors are briefed that they must feel relaxed by you. Sit as still as a statue and limit your words; a wise man hardly says anything yet reveals all! A CSA actor suggests RP (received pronounciation). Stress the first consonant of each word, explode the 'T' sound, omit your vowels and magically sound posh!

Don't pray! We all know how desperate you feel! Having observed countless GP consultation videos, mock Simulated Surgery and CSA role plays, there is one position your hands must never assume – 'the praying mantis!' This is the hands clenched together gesture which means 'please I pray, get me out of here!' Crossing arms is also very defensive. Open your hands and arms. The opposite elbow on the table is the confident 'Napoleon.'

Are you an 'ummer'? When nervous, the heart races and when it reaches 146 beats per minute, you may become petrified with fear (stage-fright), cannot utter a single word except for um repeatedly and forget even your name. Your amydala has been hijacked with intense emotions of fear and dread. What can you do? Try deep breathing exercises during your 2-minute intervals and rest period, breathe in through the nose and out through the mouth. Habituate the response by practicing role-plays. Think 'Spock' and activate your neocortex (logic and reasoning). Some advise Bach Rescue Remedy spray and others B-blockers in extreme cases.

Speak English fluently. Negative unconscious foreigner bias is created when IMGs who speak English as a second language converse in broken English, use incorrect grammar, have thick foreign dialects, mispronounce or use incorrect expressions. Here are mistakes I have identified among some Asian graduates.

'*It's Dr Khan, **here***.' Instead say '*I'm Dr Khan.*'
'***Take your** seat.*' Instead say '*Please **have a** seat.*'
'*What you have medications*?' Say '*are you taking any medication?*'
'*Is it **a** stressful*?' Instead say, '*is it stressful*?'
'*You have **got** diabetes*,' rather than '*It appears you have diabetes.*' Omit the '*got*'.
'*Do you know what that mean*?' Correct with, '*do you know what that **means**?*'
'*For how long you have had*'. Try '*how long have you had..*'
'*I'm going to **REF**-er you.*' Instead stress the second syllable and say '*May I re-**FER** you to the....*'
'*You can pick up a leaflet at **the** reception.*' Instead say, '*May I leave a leaflet at reception?*'

T must sound like 't', not 'd'. V must sound like 'v' and not 'w'.

CSA Lesson Four: The Meaning of CSA

Some sit CSA and come out of it, not knowing if they passed or not, and if they were fortunate enough to pass, they have no idea why they passed. What did they gain from revising and sitting CSA? To pass CSA, is to understand the meaning behind CSA…

The Working Party of the RCGP published a book entitled 'The Future Practitioner' in 1972 in an attempt to envision a vocational training scheme for all GPs. Prior to 1976, doctors after medical school could set up shop as GPs without sitting any exam or undertaking any specialist training. In 1976, 3-year GP vocational training began, amidst resistance to change.

The Future Practitioner book states clearly the two qualities that must not be seen in a GP trainee….arrogance and lack of empathy. So it goes without saying, that in your CSA you must project humility and empathy.

What is empathy? Empathy is simply putting yourself in someone else's shoes and echoing their thoughts and words. Empathy can be manifested in many ways…verbally and nonverbally. Verbally, a doctor may say *'oh dear'*, *'I am so sorry to hear that,'* *'are you okay?'* *'would you like a tissue?'* *'would you like me to call your husband to pick you up?'* Nonverbally, the doctor may tilt his head to one side, furrow his brows, reflect sorrow or mirror the patient's facial expressions, and finally use touch to express empathy. NB: if you are breaking bad news, it is not acceptable to weep with your patient (too much empathy!).

Ah touch you say? Perhaps why more men fail CSA then women and why patients prefer to see female GPs than their male counterparts. I often tell my candidates, that if you are unable to touch your patients, you will fail. We touch our patients when we greet them at the door with a handshake. We touch our patients when we examine them. We touch the upper or lower arm of a patient when they are sad as if to say '*I feel your pain.*' And yes, I have to qualify this as I have seen this in role-play and suggest you 'do not touch your patient below the waist (knee or lap) especially if you are a man!' I have also been asked how long to leave one's hand on a patient when demonstrating empathy? A moment's firm touch is all that is required to say '*I am sorry. I am here to support you. I am your rock.*' If a doctor is unable to communicate with touch, then a doctor is not ready to become a GP.

What else is CSA testing? Well it is asking, 'Are you a safe doctor? Can you manage risk?' The oral exam of the old MRCGP demanded you answer rapid-fire questions on a series of ethical and clinical dilemmas to test your decision-making skills under stress. I likened this to being placed in front of a firing squad. If you were answering well, the speed of questioning increased in an effort to increase your stress levels to see if your responses still made sense. And so too, the CSA is testing your decision-making and management skills to ensure you are a safe doctor even under duress. Do you admit to your patient you do not know the diagnosis, or do you make a wrong diagnosis and treat incorrectly? Do you send a patient home or refer a patient to outpatients, who should actually be admitted? Do you ask for a specialist opinion or are you too arrogant to ask for help? Do you manage the patient, rather than ask the patient how he or she would like to manage their own illness?

Why is the examiner stony-faced? I hear this a lot. GP trainers and examiners are taught to be stony-faced so as not to give away any non-verbal cues of how you are performing. Well have you ever conducted a consultation with a stony-faced human statue in the room? It is very eerie. What if we are doing very well? Look out of the corner of your eye, and you shall observe nodding, smiling and other positive cues that the examiner is happy. He is unaware he is giving away his unconscious positive affirmation that he would welcome you as a partner in his practice. If you see an examiner shuffling papers, clicking his pen, tapping his foot, playing with his iphone or looking at his watch, you know you had better improve 100% fast!

CSA Lesson Five: How to Neutralise Unconscious Bias

Amidst a complaint from BIDA (British International Doctors Association) to the RCGP that the CSA exam is racist (Pulse 25.06.09), with outcries from readers who perceive themselves as 'victims' and an application for Judicial Review of the CSA exam submitted in April 2013 by BAPIO (British Association of Physicians of Indian Origin), the fact remains that the CSA is a subjective exam. So here, I shall focus on how to neutralise negative unconscious cultural bias.

Watch a video of your consultation and turn down the volume. Just focus on observing you. Now write a list of critical comments. Be harsh and honest. Your list could consist of ....overweight, dress is too floral, unshaven, ethnic, etc. Yes I mean be brutally honest, as this is how the outside world may perceive us, sadly as societal stereotypes, including the 26 random CSA actors and examiners who have never made our acquaintance, as subjective observer bias may exist on a deeper unconscious level. My list, as a GP registrar, read...looks Chinese and not British, need to wear light make-up, voice too high-pitched and child-like, need to change voice to sound confident, accent South London/Cockney need to change to a professional educated posh accent, mispronounces 'th' as 'v', need to be careful about words with 'th', talks too fast, need to slow down, hair too long for a doctor, need to be tied up in a bun, exposed all my arms on camera, too much flesh exposed, does not look professional, need to put on a suit and cover my arms.

I know what you are thinking. But I can't lose 2 stone in 2 months. So how do we neutralise a negative unconscious bias towards obesity? Aside from job discrimination, patients take one look and are less inclined to heed your advice to take up exercise and go on a diet. You know what they are thinking, *'Why don't you take your own advice?'* The solution is to project a friendly image of Father Christmas. Put your hands on top of your abdomen, with your fingers interlocking but open, palms up. This projects a paternal image and suddenly your patients and the observing examiner are not distracted by your corpulence. In other words you have neutralised a potentially negative unconscious bias. I saw how a stout senior GP examiner transformed his image instantly just by placing his hands in this open gesture over his abdomen! It worked.

Why is it not good to wear a floral skirt or dress to an exam? Patients judge you on your attire. Society's stereotype of a NHS nurse may be based on an old English black and white film in which nurses wore dress uniforms with pinafores, white cuffs and a cap and doctors wore grey suits. They may have difficulty believing they are seeing 'the doctor' if they see you wear a floral skirt or dress. Sorry women, alas members of the public still address my husband as Dr Coales as it is unconsciously easier to believe a man is a doctor than a woman. So for the CSA exam and in reality, wear a black or navy blue conservative skirt suit to project professionalism and make a clear statement that you are the doctor. The public wouldn't expect a female airline pilot to show up to work in a dress but expect to see her wearing a suit and preferably a trouser suit to distinguish herself from the stewardesses in skirt suits. The public wouldn't expect a female army officer to dress in a floral dress at a ministry of defence meeting but in a skirt suit. A patient declared to me that he wished to see me again because.... I was professional (I wore a suit)! Members of the public also like perceived value for money. They pay £100 for a 20-minute consultation with a private GP, expect him to be wearing a tailor-made suit with impeccable manners and think they are getting a better service. Irrational but there you have it, public (mis) perception and especially relevant in the context of revalidation incorporating patient satisfaction surveys.

If you sport a beard or moustache, my advice is to shave it off just for the exam, unless it is for religious reasons, ie you are Sikh. It will grow back. If Sikh, instead keep it trim. What subconscious image does it project when a professional sports facial hair...unclean, deceitful, hiding a weak jaw-line? In other words, nothing positive. As women are more judgemental than men and examiners are also invited from the Ministry of Defence, present a clean shaven professional face. Safer to please a larger audience.

Grooming is essential! Yes and that includes removing nose piercings, multiple ear piercings, taking a shower to remove body odour and using mouth wash if you have eaten something too spicy or with garlic or onions. I've seen it all. Men try a dash of cologne!

Ethnicity. Ah now we come to the sensitive topic of 'race'. Black IMGs are quoted to have a CSA pass rate of 35% and Asian IMGs of 36.9% in the 2010-11 RCGP report. Having met numerous Nigerian GP trainees in my courses, I find their English fluent and they go on to pass easily by correcting the 'th' pronounciation. Some Asians are hampered by language and for me as a teacher, the male NEFL (non-English first language) Asian candidate is the most challenging to help clear this exam. If a male NEFL Asian is based in Scotland or Wales, I ask him to focus on emphasizing the lyrical Scottish or Welsh accent he may pick up at work so the immediate first impression is of a local graduate and not that of a 'foreigner'. An immediate first impression that the candidate before the observer (actor and examiner) is foreign, may elicit negative unconscious bias. One examiner even advised his Asian trainee to limit his words so as not to give away he was foreign! A Chinese IMG (RCGP pass rate is only 25%) got rid of his Cantonese accent and passed his final CSA exam.

My suggestions are designed to get you through the CSA by neutralising negative unconscious bias. Be aware of what may be projected to the outside world of strangers in a country in which discrimination sadly still exists, whether overtly or covertly, so that you are not pre-judged but are assessed purely on your GP skills. This is about helping you clear 'your driving test'. They're not about changing who you are. Who doesn't smarten up before a job interview? Sadly CSA has become a career-breaker for many, and this is what this book is trying to address to help GP trainees.

CSA Lesson Six: The Power of Words

We have just looked at how 'visual perceptions' affect marking of a subjective CSA exam. Now we turn our attention to how our words are influencing both our patient consultation and our CSA exam outcome.

'The Power of Words' by Dr Bernie Siegel talks about how the power of doctors' words may impact patients' lives. *'Parents, teachers, clergy and physicians have the ability to change lives with their words.'*

Frank Luntz, a political pollster and author of bestseller "Words That Work: It's Not What You Say, it's What People Hear" also discusses the impact of words and how to use words with tact.

So let us reflect on how CSA candidates use language. A textbook scenario honed into every doctor during medical school is the concept of 'breaking bad news.' A standard phrase may be *'I'm afraid I have bad news Mr X. The chest x-ray shows you have a sinister shadow in your lung.'*

Now what does the patient actually hear? Let us dissect this phrase. *'I'm afraid'* may be a condescending phrase and has connotations of the sales ladies telling Julie Roberts' character, Vivian, in the film 'Pretty Woman, '(*I'm afraid), I don't think we have anything for you. Please leave.*' 'I'm afraid' also implies unconsciously that you as a doctor are fearful, unsure of yourself, hesitant. Furthermore, neurolinguistic programming (NLP) suggests you are influencing the listener to feel fear. The classic example used in NLP training is the phrase 'Don't think of pink elephants.' So the class recalls only the last few words (NLP) and conjure up thoughts of a pink cartoon elephant. So here the last word heard was 'afraid' and your patient starts sweating. His heart starts racing. His face becomes petrified and eyes glaze over.

'I have bad news.' Oh dear, now you have declared yourself the harbinger of doom, the Grim Reaper. The examiner and actor hear

the words '*I*' '*bad*' in the same sentence and are influenced to fail you as you have just declared 'I (am) bad' in front of the examiner. Poor patient is now thinking you are Shipman 'I (am) bad' or he himself is 'bad' or 'evil' and has brought cancer upon himself for being a bad person.

On the verge of a heart attack, you now tell him '*you have a sinister shadow.*' The word '*sinister*' can appear as '*sinestre*' in French for '*left*' and is also derived from the Latin '*sinestra*' meaning '*from the left-handed side*', devious, cunning, presaging menace. So you have suggested subliminally in front of the examiner and actor patient that you are a '*sinister doctor*', Shipman perhaps? Now combine this with the word '*shadow*', and the patient will now burst into tears if female or stand up and get ready to punch you in the jaw, if male. 'Shadow' and 'fear' conjure up Psalm 23 '*Ye though I walk through the shadow of death, I will fear no evil.*' You have effectively influenced your patient to prepare for his own funeral. No wonder male patients get violent. I think that qualifies as a 'fail' your CSA. The answer is to be honest and say 'cancer', but deliver with empathy and compassion.

Better to give a warning shot and use the empathetic phrase '*I am sorry.*' '*It's not good news. I am sorry; the results have come back as cancer.*' And this is where human touch is so important to convey care, empathy and mutual understanding of human anguish and sorrow. Touch the patient's arm to demonstrate empathy and if you are seated too far away, move your chair or crouch down and kneel by your patient to touch. Humility = humanity.

Another phrase I hear frequently in role-play is '*Don't worry.*' NLP suggests that you have just told your patient to 'worry' and surely, they start worrying, fretting, questioning, and experiencing palpitations. Best to use positive affirmations to assuage anxiety and say, '*Everything will be fine.*' This conjures up emotions from the 19th Century Camberwell-born English poet Robert Browning, '*God's in His Heaven; all's right with the world.*' The only caveat is if the patient has metastatic cancer, then we avoid false hope. Instead we say 'we will do the best that we can.' For prognosis in early detection try '*it's early days now.*'

Here is a caveat for overseas-trained candidates who speak English as a second language. If you utter one sentence in 'broken' English you may receive 5/9 or fail for that station. It is not the 'race' card that is being selected but simply the fact that you are sitting an English postgraduate medical licensing examination. The ability to speak fluent English is vital to prove in a medical licensing exam that you will be a safe doctor practicing in the UK. So if this is your umpteenth re-sit, invest in linguistic lessons to deliver in flawless English, '*spoken like a true gentleman*'. Many of you will argue that you have passed the IELTS English language test. Think, who are your patients in this exam? In CSA, they are British actors, many of whom are graduates of the Royal Central School of Speech and Drama, who will either help or hinder you if influenced by unconscious bias. They will subconsciously compare their command of the English language and delivery with yours. They are not your typical NHS patient, many of whom you may converse with in Urdu on a daily basis! Try paraphrasing. Replace '*do you have any fever,*' with '*any fever?*' With one month to perfect your English, use this '6 words or less' K.I.S.S. technique.

Now turn up the volume on your video playback and transcribe the words you use to communicate with your patients. What exactly do they hear? Remember to strike the word 'bad' from your vocabulary. Help the examiner pass you. Do not contribute to your own CSA demise! Why not incorporate the word 'excellent' at any opportune moment? And next time you ask the actor to 'tell me a **bit** more' you will understand why the actor only gives you a snippet of a clue. What's in a name? To neutralise unconscious foreign name bias, introduce yourself using an Anglicised version of your name. Dr Debashish can be Dr Deb. From November 2012, full names are posted on respective CSA room doors.

CSA Lesson Seven: The Willing Suspension of Disbelief

Now that you have mastered how to minimise conscious and unconscious perception bias of appearance, cultural differences and speech, let us focus on the rest of the CSA exam, the actors, the examiners and the setting for a complete 360 ° assessment of CSA.

The actors are NOT real patients. The FRCS, MRCS, MRCOG, MRCP, MRCPCH and DCH postgraduate medical exams use real patients for their examinations. The MRCGP employs actors. Does this make a difference? Absolutely! When an actor presents with red eyes, do not assume this is the case before you. It simply means the actor did not sleep well the night before. He is not wearing customised red contact lens to mimic red eyes. Let's try again.

When you perform a BP on an actor and the BP reading is high, this is NOT the case being tested. The actor cannot mimic high BP, and nor was he chosen for the role play because he happened to have an incidental finding of high blood pressure. So what do you do? You may hope that the actor says '*That is not required for this station*,' that the actor or examiner gives you a card with the correct BP reading for the station or you may remain flummoxed as you feel the actor should be advised to see his own GP for further discussion of his BP and management.

When an actor patient presents with a history of thyroid disease but has no goitre or exopthalmos when you glance at your healthy actor patient, remember she may still have 'a goitre and bulging eyes.' So ask permission to examine her and hope you are handed a card which confirms or refutes any imaginary physical findings. Do not rely on your visual senses. If the actor yawns in the afternoon, he may not be role-playing as hypothyroid, but may genuinely be tired role-playing the same case 26 times!

When an actor patient presents with a history of PCOS and yet is not obese, hirsute or sporting acne, remember to ask to examine the patient and hope that a card will reveal whether the patient is obese with facial hair and/or acne. The difficulty is that your senses, intuition, and experience tell you that the patient before you does not have PCOS but that you are supposed to 'pretend' that the actor does, even though the actor knows nothing about this condition, save what she was briefed that morning. This is where the most experienced clinicians face difficulty with CSA. They rely on years of experience of assessing a patient's appearance (unwell vs. well), manner and demeanour, skin pallor, icteric sclerae, quality and strength of pulse, temperature of skin on touch, quality and rate of speech combined with respirations, use of accessory muscles of respiration, and so on and so forth and combine all these cues to help reach a working diagnosis. All these clinician tools are absent in the CSA when using healthy actors.

When an actor patient discusses knee pain and yet you glance at his knee and it is perfectly shaped, remember to ask to examine the knee and hope to receive a photo image or a note on your ipad indicating an abnormal knee. The difficulty may then be to examine the abnormal knee in the image with an actor and his normal knee. You cannot elicit knee clicking, popping or crepitus. I suggest using the 'willing suspension of disbelief'.

When booking, you may only choose a handful of sessions to avoid. Avoid the PM sessions, especially on a Saturday. The actors and examiners arrive early to be briefed and rehearse 1 role-play to be enacted for the entire day. That is 26 times of role-playing the same case for one actor (13 times in the morning and 13 times in the afternoon), and it is 26 times for one examiner to observe and mark the same case over and over again. And he may examine 3 days in a row! You got it! Acting and marking fatigue. Can we really rely on the actor remembering every sentence of his script on the 26th take of the role-play or the examiner to remain attentive, the 26th time he observes the same case? Better not take any chances, avoid the afternoon PM session! Also those who attend in the afternoon, I cordoned off for 2 hours in a holding bay prior to commencement of the exam. Bring food and books to read!

What about choice of day for your CSA? Well how would you feel at the end of a week of examining or acting to then be told you had to come into work on a Saturday? Who likes to work on a Saturday? Better hope for a weekday AM CSA session.

Perform a basic GP exam. Some stations assess your clinical examination skills. Then how is it that GPs with years of surgical or medical experience, MRCS or MRCP, are failing their CSA? What was tip number one? The actors are NOT real patients. So when a GP with MRCS conducts a proper shoulder exam, with the Neer test, Apley scratch test, Hawkins test, empty can test, etc., the actor can only look baffled, as he has not been briefed how to respond other than to a basic range of movements of the shoulder that he learned that very morning. After your exceedingly thorough examination of his shoulder, you too are perplexed, unable to distinguish between supraspinatus tendinopathy, rotator cuff tear, etc., as the physical finding responses given by the actor may appear nonsensical. Stick to a GP exam, and discard your specialist knowledge for 1 day.

Read the free RCGP Curriculum online which lists all the cases that may be tested. Cases discussed above are from their respective curricula under learning outcomes for women's health, rheumatology/ musculoskeletal and metabolic problems.

CSA Lesson Eight: Practicalities

From November 2012, the CSA exam has been conducted at the RCGP headquarters in Euston.

Pack all your equipment and copy of your British National Formulary the night before, much like preparing for an expectant baby. Have your CSA bag ready to go. The list of equipment required is on the RCGP website. You would be amazed that some doctors have forgotten their BNF or stethoscope in a panic to get to Euston. Do not invest in a fancy doctor's bag as you will be supplied a clear plastic bag to carry your belongings on the day and a locker for item storage.

Eat porridge for breakfast, as no food or drink is allowed in the rooms; you may bring food for the holding area for PM sessions. If you are a diabetic, please inform the invigilator who will allow an exception for food. The old adage of porridge in boarding schools applies here, as it has a low glycaemic index and allows you to sustain energy levels over a long period.

When you arrive, you may be waiting in a cordoned room with other candidates, while the actors are being briefed and examiners are preparing. You are permitted to bring reading material for the waiting room.

You are allocated one room on one floor of the three examining floors. The examiner and actor walk from room to room as a pair. So when you enter your assigned room, quickly empty out your equipment and set up the room. When the invigilator checks on you, ask that the blinds/curtains be drawn and that you have a heater for the room if the room is cold. Ensure your room has a working clock.

The cases differ from day to day and week to week. If you are a re-sit and come across a familiar case, note that there may be permutations of the same case. You may be assessed on a completely different focus or angle than before, so be flexible.

Each room is equipped with a large digital count-down desk clock. Aim to complete in 8 minutes instead of 10, as the CSA is assessing whether you are ready for partnership, to be released into the world as a Future Practitioner and not as a GP registrar sitting CSA. In real-life, one is also required to type up a medico-legal report after seeing the patient within your 10-minute consultation and complete a physical examination.

If you do not complete in time and the bell goes, the actor will 'morph' into a zombie. He will stand up, turn his back on you and exit the room in silence. It is a very spooky experience, as real patients offer their 'thank you's' and 'goodbyes' before exiting your room. This is another reason to try your best to complete your consultations on time.

If an actor tells you 'that is not required for this station', stop examining him! It means the station is not testing your examination skills, but rather another domain.

If a case requires a chaperone, ask for a chaperone or you will fail. If the actor is playing a patient with a learning disability, ask permission for a 'nurse' to help you with the exam as 'chaperone' would then be jargon. If the examiner does not step forward as a chaperone, follow the GMC guidance on chaperones and assess without a chaperone, only in an emergency situation (fear of cancer, suspected testicular torsion, etc.)

The first two rooms are equipped with video camera recording. So during your 2-minute interval between cases, please remember you are still being observed…no funny faces, no burying your head in your arms in exasperation, etc. The advantage of being allocated such a room, is that if there is any impropriety, i.e. the actor walking out responding by mistake to the bell on another floor of the building, you may inform the invigilator on the day with hard evidence. Be aware that tapes are erased after 2-3 days, so best to lodge your complaint on the day and not weeks after.

Free re-sittings of CSA have been offered to those who report to the marshall on the day of any CSA exam disruptions i.e. fire alarms going off and being asked to leave the building during the exam.

If you require the loo, you will be accompanied by a 'warden', so limit coffee or tea on the day. In fact, avoid all stimulants, as the exam itself is giving you more than enough of an adrenaline rush and it is vital to keep calm to perform at your optimum and to 'ease and reassure' the CSA actor patients.

For those of you who are facing re-sits, you may email exams@rcgp.org.uk and ask to make a data subject access request for all your past examiners' marking sheets. This will be supplied within 40 days and is free of charge. The remarking charge of £60 or a formal appeal of £800 has rarely resulted in a change to the marks.

Una Coales's Consultation Model

The story behind the 10 CSA Commandments

Occasionally I would have candidates come to me and tell me they were sitting their CSA the very next day! And so I came up with the 10 CSA commandments, a simplified consultation model that they could learn over-night to face any one of the now 650 CSA cases in the bank. I call it the 10 commandments to remind the GP trainee that we GPs have taken over the pastoral role of clergy. In smaller communities, people would off-load to their parish priest in confidence, but now in a fast-paced society, the GP has become the first port of call. CSA I tell my GPs stands not only for clinical skills assessment but also for Caring, Saintly Angel. It reminds us how humble and privileged we are to receive private confidences from patients and the immediate trust they have in us to help them with their problems. It reminds us of the 'Golden Rule' inherent in any and all religions across the world, 'Do unto others, as you would have them do unto you.' It reminds us to be kind and caring, 'cum scientia caritas', the very motto of the Royal College of General Practitioners. And finally it reminds us of the teachings of the 'Future Practitioner' written by the working party of the RCGP in 1972 in which we hear of the 2 qualities we must never see in a GP registrar....arrogance and lack of empathy. And so for the CSA, we demonstrate the exact opposite – humility and ooze empathy. And if all else fails, it reminds us to pray to God before our exam. God helps those who help themselves!

The reason why my consultation model is so simple and effective is because it is aimed at the level of a lay actor or patient, i.e. 100% patient-centred. An ancient Greek tutor came by to provide lessons for my daughter. She glanced at the 10 commandments on the white board and asked me, '*are you teaching doctors how to speak English?*' She herself was a Cambridge postgraduate with a master's degree in ancient languages.

Often times, we, as postgraduate professionals, forget and assume our patients likewise have similar university and post-graduate medical degrees and may converse as professional colleagues. They do not. So if your child cannot understand you, then neither can your patients. Keep your words short and simple. Remember the acronym K.I.S.S! The more you speak, the more doctor-centred the consultation becomes and the less likely you are to pass. Less is more. And I know how much you love describing the pathophysiology of a patient's condition to him but have a look at his non-verbal body language. Are his arms crossed, is he dozing off, are his eyes glazed or wandering around the room? Yes, keep even your explanation of the disease short and simple.

I equate the philosophy behind the 10 CSA commandments consultation model to being a wise holy man who hardly speaks, yet reveals all. And speaking of the clergy, ensure your tone of voice is soft and gentle at all times! Transform your consulting room into a haven of peace and tranquility for you and your patient. Bring an element of calmness to your patient's hectic life.

Why do I bring up clergy and not a policeman or teacher? 26 subjective judging examiners and actors from all walks of life will all agree that the most caring person in the world is a nun or a priest, and not a policeman or teacher. We have to remember that the 'judges' all come with preconceived schemas. An examiner or actor may have been issued with a speeding ticket or parking fine so may be unconsciously negatively biased towards a doctor who speaks abruptly like a policeman. An actor may have had a harsh teacher at school so we do not want to appear like an admonishing teacher or a teacher who lectures the entire 10 minutes. So our safest option is to think 'Vatican City priest or nun' in hopes we do not inadvertently activate a pre-determined schema in our audience of 26 random judges. As an aside the hijab unintentionally conjures up unconscious positive bias. An actor is less likely to be confrontational with a Muslim female GP trainee wearing a black or white humble religious hijab.

Una Coales's Consultation Model

The 10 CSA Commandments

1. ***'How may I help you? Tell me more about X?'***
   Here we make it a point to use *'may I'* and not *'can I'*, as English grammar dictates we say *'may'* and not *'can'*. The word *'help'* has wonderful NLP tones and invites the patient to take charge of his/her consultation. We are placing ourselves as clinicians in a humble, meek, and serving capacity. Patients nowadays see themselves as NHS consumers so naturally we adapt our language to provide the proverbial American *'customer service with a smile.'* Ideally we would prefer if patients came to us seeking professional medical advice rather than present a shopping list of 'wants' rather than 'health needs'.

   'Tell me more about...' allows the patient to speak and share his own narrative. In fact we want the patient to speak for at least 90 seconds. However, in the CSA, actor patients are very monosyllabic, so the occasional prompt may be necessary... *'what happened next?'* The phrase *'tell me more about'* also gives an air of timelessness, as though the GP has all the time in the world and would like to know everything about the patient in front of him. The patient has, in effect, become the centre of the GP's universe for 10 minutes, but time is relative and with my consultation model, the patient will feel as though you have stretched time itself!

2. ***'What's worrying you about X? Is there something at the back of your mind? Has anything happened/ changed recently?'*** Here these are three prompting questions to ascertain the patient's hidden agenda. Twelve times out of thirteen cases, the actor is hiding his true concern so will need non-verbal and verbal prompting to allow the actor

patient to reveal his deep underlying concern. You will notice I do not use Pendleton's words 'ideas, concerns and expectations' when speaking directly to a patient. If you ask a patient, *'what do you think?'* the usual sharp retort is, *'you're the doctor; you tell me!'* The words *'concern'* and *'expectations'* in my opinion are too alienating to a young patient who may have left school early or in cases where English is not the patient's first language. Again we keep our language simple to facilitate mutual understanding and avoid being perceived as too doctor-centred or intellectually arrogant.

3. **Rule out red flags. Ask family history and medication only if it is relevant**. Here candidates vary in their history taking. Some over-run, determined to pass an 'MRCP' exam and forget they are sitting an 'MRCGP' exam in which the ability to time-keep to 10 minutes is an art in itself. Keep your history-taking concise and relevant to ruling out the red flags and helping make a working diagnosis. Think airline pilot and not medical student.

4. **Background information.** *'Do you work?' 'Self-employed?' 'Who's at home?' 'How's your marriage?' 'Finances?' 'Sports?' 'How does X affect your ADL?' 'Stresses?' 'Do you smoke or drink?' 'Have you thought about cutting down or quitting?'* You are putting the disease in the psychosocial context/setting. This will allow you to adjust your management plan to accommodate the patient's lifestyle and beliefs. The underlying concern may be social, ie a need to get back to work as soon as possible if the actor patient is self-employed as a SC1 certificate only covers NI payments.

5. *'May I examine you please?' 'I'd like to listen to....'* Here either the actor or the examiner will give you a card or ipad image. You may be given a photograph or just a

few words describing the findings. Remember to offer a chaperone for any intimate examination but respect the patient's desire for none to be present if she so wishes.

6. *'I know you were worried about X. Based on a, b and c, you do not have X, you have (diagnosis).'* Each station has a diagnosis so do remember to say what you think the diagnosis or working diagnosis is. You can't complete a consultation and never tell your patient what is wrong with her? So often I hear candidates say, *'oh I never actually said what the diagnosis was'* and wonder why they are a re-sit. Part of a licensing exam is the ability to reach the correct diagnosis with the information provided or gleaned from your consultation.

7. **Explain the diagnosis and check patient's understanding. *'Do you know what (diagnosis) means?'*** Never assume that your patient does not already know. He might have read up on the condition and be more of an expert than you! Keep your explanations brief and use visual aids if necessary. Some conditions are more easily explained with a diagram. GPs often use patient leaflets as aids knowing that a detailed explanation could take a full 10 minutes by itself! Show risk with the BNF chart.

8. **Reassurance and hope. *'It will get better. There are lots of things we can do. The options are a, b or c. What would you like to do?'*** Here it is vital to share the management options with the patient and allow the patient to decide, i.e. make an informed decision. If he cannot, then be patient and repeat or simplify the options. Keep your options short. If the patient prompts for more details, then you may elaborate. A customer in a restaurant does not expect a waiter to draw up a list of options and then proceed to explain how each item is prepared?

9. **PILS.** *'Would you like a patient leaflet on (diagnosis)?'*
    Take a few moments to explain what may be found on the
    leaflet. *'May I leave a leaflet for you at reception?'* or you
    could jot a discreet note for the receptionist to print out a
    leaflet on HIV testing, showing patient sensitivity.

    If you do choose to write a prescription, ensure you
    prescribe generically, fill in all the boxes and make sure
    the writing is legible. Cover potential side-effects of
    medication such as carbamazepine. Else say *'May I start
    you on aspirin? May I leave you a prescription for 300 mg
    of aspirin to be taken once a day, at reception for you?'*

    **Roger Neighbour's safety-net and appropriate follow-
    up.** *'Please come back and see me if you are not better or
    you get worse.'* *'May I see you next week for the test
    results?'* *'If you have not received an urgent breast clinic
    appointment within 2 weeks, please let me know.'* Alert
    the patient to potential red flags so the patient is also aware
    of the dangers to look out for. Think preventative
    medicine. Make some suggestions relating to work-
    adjustments, ergonomics, etc.

10. **Check patient's final understanding** especially if you are
    told the actor is in his or her teens or 20s. This will be a
    clue that the actor patient may have poor understanding
    and his or her final understanding needs to be checked to
    be safe. *'Could you tell me what you have understood
    from today?'* *'Have I reassured you completely?'* *'Thank
    you Mr. X.'*

11. Buzz words: CSA actors are primed to divulge based on
    certain triggers. If you notice an actor with a limp,
    comment on the limp and he or she will start revealing
    clues. If you need to take an STD history, begin by saying
    it is 'confidential'. Look out for verbal and nonverbal cues.

Sample Dialogue for CSA

Good morning, Mr. Jones. How do you do? My name is Dr X. Shake hands. Please come on through and have a seat.

How may I help? What can I do for you today? What seems to be the matter?

What do you mean by 'dizzy'? May I ask you more about that?

Ask red flag questions to exclude sinister condition.

Do you have any idea as to what is going on? What do you think it could be? What are you worried about? What does your husband think is going on? What were you hoping we could do for you today?

Do you know anyone who has HIV? Cancer? Had a heart attack? What is your concern? Is there anything in the back of your mind you are worried about? Has anything changed recently?

Do you have a family h/o cancer/ heart attacks/ bowel problems/ diabetes/ depression? (Only if it is relevant).

You mentioned concern over...Had you any worries when you went for the x-ray?

You look sad/ stressed. You seem low. Are you coping? Are there any stresses in your life?

You mentioned earlier that you felt low; could you tell me more. Is there something else going on?

Any financial worries? How are your finances? What do you do for a living? Are you self-employed? Are you the sole earner?

Is it affecting your work? How are you managing at work? Are you managing alright at work? Do you enjoy your job?

If you were in a different job, how would you feel? What did you do before you retired?

How do you spend your day? What do you do for fun? Hobbies? Sports?

How is x affecting your day? (Activities of daily living)

Who's at home with you? Any family or friends nearby?

Are you married? How is your marriage? (Domestic violence?) Do you have children? (At risk?)

May I ask you some questions about your general health?

Do you smoke? Have you thought about quitting? Are you ready to quit? When you are ready, I can refer you to our stop smoking clinical. If no, say 'excellent.'
Do you drink? How much? 21u M 14u F Have you thought about cutting down?
When was your last pap smear? Have you had your flu jab this winter? (Only for asthmatics and elderly).
Just a routine question. Do you use drugs? (Only for high risk or teenagers).

May I examine you please?
I'd like to look at the back of your eyes, check the nerves to your face and check your blood pressure.
I checked you over and I think you are right. I think you have 'a common cold'.
Do you know what anaemia means?
What is your understanding of diabetes?
I checked you over and I know you were worried you may have a brain tumour, your brain is fine, you have had a x.
Does this make sense to you?

Let's share the management options together. The options are:
We can do nothing and wait and see (depression - exercise, self-help books also).
We can arrange a blood test to rule out anaemia.
You can try OTC x or shall I write you a rx for something stronger?
Would you like me to arrange some counselling (practice, psychologist)? Would you like a practice leaflet on x?

Referral for physio/ 2nd opinion/ hospital specialist/ etc. Would you like to be referred straight away or try something else first?
Would you like to see the diabetic nurse for further education?

In the meantime, would you like to take time off for depression?
Have you heard of a self certificate? SC 2 form for common colds.

Which would you prefer? What would you like to do? Can you repeat back to me what the plan is?

Would you like a leaflet on x, and if you have any questions, please come back and see me? (Internet/ self-help books/ local support groups)

How about seeing our practice benefits advisor to help you with your financial worries as you are self-employed? Have you been to the Citizen's advice bureau (debt)? Have you met with your bank manager?

Write rx and explain how to take medication and its possible side-effects. Can you repeat back to me how you will take the medicine? What will you do if the antibiotics don't work and you get worse? (Consider a delayed antibiotic prescription if the actor patient insists on antibiotics for a viral illness).

Could you make an appt to see me in a week for...pap smear (25-64), review, MSU results, flu jab (>65yo), etc.? See me sooner if you don't get better or have (red flags).

So what have we agreed to do? To make sure you have understood, can you repeat back to me what we have discussed. What will you tell your husband when you get home? What have you understood from today? Have I reassured you completely? Goodbye Mr. Y. Get up, shake hands 'a pleasure to have met you' and show Mr. Y to the door. Manners!

CSA Marking Criteria
(source RCGP CSA Feedback on www.rcgp.org.uk)

<u>RCGP: A guide to the new MRCGP Assessment Process Practical aspects for trainers and course organiser</u>: CSA domains, blueprint and theme headings.

<u>RCGP GP Curriculum for case selection</u>: CV, Respiratory, Musculo-skeletal, Neurology, Eyes/ENT/Skin, M/F Health, Renal, GI, Genetics, Learning Disabilities, Acutely Ill, Healthy People, Care of Elderly/ Cancer and Palliative Care/ Drugs+EtOH/MH.

<u>Assess</u>: acute and chronic illness, undifferentiated illness, psychosocial, preventive lifestyle, other.

<u>I. Data-gathering</u>, technical and assessment skills: Gather & use data for clinical judgement, choice of exam, ix's & their interpretation. Demonstrate proficiency in performing exams & using diagnostic and therapeutic instruments.

<u>II. Clinical mx skills</u>: Recognise & mx of common medical conditions in 1° care. Demonstrate a structured & flexible approach to decision-making. Demonstrate the ability to deal with multiple complaints and co-morbidity. Demonstrating the ability to promote a positive approach to health.

<u>III. Interpersonal skills</u>: Demonstrate the use of recognised communication techniques to understand the pt's illness experience and develop a shared approach to managing probs. Practise ethically with respect for equality and diversity, in line with the accepted codes of professional conduct.

BORDERLINE GROUP METHOD
FOR MARKING BASED ON POINTS (FROM SEPT 2010)

DATA GATHERING

Organised and systematic in gathering info from hx, exam, ix.

Identifies abnormal findings or results and implications.

Data gathering guided by probability of disease.

Undertakes exam competently and uses instruments proficiently.

CLINICAL MANAGEMENT

Makes an appropriate diagnosis .

Develops a mx plan (including prescribing and referral that is appropriate and in line with current best practice .

Follow-up arrangements and safety-netting are adequate.

Demonstrates an awareness of risk and health promotion.

INTERPERSONAL SKILLS

Identifies the patient's health beliefs, agenda and preferences. Makes use of Verbal and Nonverbal cues.

Develops a shared management plan or clarifies the role of doctor and patient.

Uses explanations that are relevant and understandable to the patient.

Shows sensitivity for the patient's feelings in all aspects of the consultation including exam.

GLOBAL

Organised and structured consultation

Recognises the challenge

Good time management

Appropriate doctor-centredness

<u>CSA Examiner Feedback (source: www.rcgp.org.uk)</u>

1. <u>Data Gathering</u> Organise and systematic in history taking, exam and investigations to develop a mx plan. Be appropriately selective in the questions you ask, the tests you request and the exams you choose to undertake. You may feel that it would be better to be "on the safe side" by ordering a battery of tests and whilst understandable, this is not good practice and will make you appear indiscriminate. Likewise, hx taking and exam is not expected to be all-inclusive and should be tailored to the circumstances and include psychosocial factors where relevant. Explain to the patient what you are doing and why. This is good for patient care and will also demonstrate that you have a clear and systematic approach. Explain exactly what tests (e.g. blood tests, if appropriate) are going to be necessary for further pt mx.

2. <u>Identify abnormal findings or results and recognise their implications</u> Demonstrate an ability to identify significant findings and act on them appropriately. Issues identified may need to be prioritised. The ability to manage uncertainty and risk is important. This is clinical rather than IPS and requires you to make sure that you can correctly interpret the significance of test results or the findings of physical and mental state exams. Pay close attention to

your ability to assess risk and to pick up on abnormal findings and deal with them safely (safety-net).

3. Data gathering appears to be guided by the probabilities of disease You should try to apply your theoretical knowledge to the clinical situation and be appropriately selective in your choice of enquiry and/or exam. The skill here is to be selective and to demonstrate that you understand what is likely, what is less likely and what is unlikely but important. Improve your skills by explaining your approach to the patient. For example, explain what you are looking for, what you think the likely diagnosis will be and (where appropriate) what you feel is unlikely but needs to be ruled out.

4. Undertake physical exam competently, or use instruments proficiently - demonstrate the appropriate and fluent use of instruments. Improving these skills is a matter of practice and it pays to spend time developing a systematic method that you can practise over and over again to become fluent, appear competent and confident.

5. Clinical management Make appropriate diagnosis. You should consider common conditions in the differential. Making a diagnosis means committing yourself on the basis of the info you have available to you. When you have made a diagnosis in consultation, state this clearly and explain to the patient. If your summary is vague, it may not be clear that you have made a diagnosis.

6. Develop a management plan (include prescribing and referral) that is appropriate and in line with current best practice.

7. Follow-up arrangements and safety netting are adequate Your management plan and follow-up arrangements should reflect the natural history of the condition, be appropriate to the level of risk and be coherent and feasible. Possible risks and benefits of

different approaches including prescribing need to be clearly identified and discussed.

8. Demonstrate an awareness of management of risk and health promotion In order to manage risk appropriately, you should make the patient aware of the relative risks of different approaches. Health promotion requires doctors to demonstrate an awareness of health (not just illness) and be proactive in maintaining the patient's health. Managing risk and living with uncertainty are key skills in general practice. Your knowledge base is important here, as is your ability to integrate that knowledge with the specific info you have gained about the pt. Try to be aware of health promotion issues and apply these appropriately. The use of computer-generated prompts can be helpful.

9. Interpersonal skills. Identify patient's agenda, health beliefs & preferences / make use of verbal & non-verbal cues. Show competence in using listening skills to identify the patient's agenda, health beliefs or preferences (patient-centred).

10. Develop a shared management plan or clarify the roles of doctor and patient. This may be improved by responding appropriately to the patient's agenda and by attempting to involve patients in making decisions regarding their prob. Clarify the respective roles may involve reaching agreement with the pt as to what will happen next, who does what and when and the conditions (i.e. the timescale and circumstances) for FU. There should be a shared understanding before the patient leaves and this can be confirmed by asking the patient to summarise what they have understood.

11. Use explanations that are relevant and understandable to the patient Avoid the use of jargon, to establish the patient's health beliefs and tailor your explanation to these. Whether or not your explanation has been understood can be checked through non-

verbal communication but also (and more explicitly) by asking the patient to summarise.

12. Show sensitivity for the patient's feelings in all aspects of the consultation including exam.
Demonstrating interest in and warmth toward the patient and seeking consent for any clinical exam is important.

13. Global Organised / structured consultation. Assessors feel that the consultation flowed, for ex that the tasks of the consultation were sufficiently integrated. Use a consultation model. Use skills i.e. explaining what you are doing and summarising at appropriate times can help to demonstrate a fluent approach.

14. Recognise the challenge (e.g. the patient's problem, ethical dilemma etc.) Identify the patient's problem/agenda or the challenges and appropriate priorities from the doctor's perspective. Be alert to verbal and non-verbal cues. Encourage the patient to share his/her thoughts and expectations.

15. Show good time management – consult effectively and efficiently in 10 minutes. Ensure you remain focussed on the problem.

16. Show appropriate doctor-centeredness – spend time on encouraging and assisting the patient to contribute to a shared dialogue between doctor and patient. Being doctor-centred is appropriate for certain tasks i.e. when using closed questions to take a clinical history. The doctor's agenda i.e. gathering data for health promotion, is important, but the challenge is to achieve an appropriate balance with the patient's agenda. Practising listening skills and being alert to verbal and non-verbal cues might help develop a more patient-centred style. Patient-centred doctors are responsive to patient preferences and work to develop common ground and a shared understanding.

<u>Advice from a successful CSA candidate:</u>

'*With regards to the exam this is what I have to say..*
1) The "10 commandments" are DEFINITELY gospel!.PRACTICE,
PRACTICE, PRACTICE!!!! PRACTICE AGAIN & AGAIN...&
AGAIN with friends & colleagues to get into the habit as it were
and so that doing these types of consultations almost becomes
second-nature.

2) Yes it is important to check understanding (at least try to), but it
doesn't necessarily have to be at the end ...in one station for
example where I thought I was running out of time, I sort of
devised a way to do this briefly by asking the patient once or
twice.."so are you with me so far?" and in another where I realised
asking him to feedback to me will take too much time, I asked him
"may I just ask if everything's clear or if you need me to clarify
anything?" However if one obviously has got enough time, then of
course, definitely ask them to feedback to you what they've
understood   from   the   whole   consultation.

3) BE VERY POLITE and ALWAYS welcome with a smile!...my
trainer told me after observing my consultation in the surgery, that
he believed my "charm" towards my patients would stand me in
good stead during the exam...and I believe it did!....in one case, the
patient walked into my consulting room & rebuffed my
outstretched arm for a handshake, I didn't let that perturb me, I
simply kept my smile on and graciously asked him to take a seat
and   did   a   VERY   patient-centered   consultation.

Similarly in the case of the angry patient who basically just really
tried to get under my skin...I just remained VERY apologetic, VERY
sympathetic & VERY empathetic throughout while outlining what I
would do to address his situation...I am so sure of it that they are
DEFINITELY looking for the "Vatican nun" GP as Dr Coales says
and it is important to "ooze" compassion from every pore really.
4) If and when appropriate offer a leaflet but say what the leaflet
will do/contains i.e. "may I offer you a leaflet on....? it contains

more detailed information about the condition or about your condition or outlines exercises you can do…

5) Try to NEGOTIATE with patients who have their own agenda. They may be difficult and unyielding at first but if you remain focused with a neutral tone of voice and try to make them consider what you are offering instead, they may acquiesce. However if they do not, DON'T force the issue, give them what they want and also offer/arrange a further appt to further discuss things..

Perhaps MOST IMPORTANTLY is to try to relax and remain calm & collected..very difficult I know, as it is an exam but really that's the BEST thing one can do for oneself. Getting all flustered and panicky before the exam or on the day would MOST LIKELY affect one's performance .If a station didn't go too well, FORGET about it IMMEDIATELY and RE-FOCUS your mind to the NEXT one...VERY, VERY, important...even if one failed that station (most times we tend to assume we performed worse than we actually did); it would be a disaster if one allowed the despair or bad feeling about that station to affect the remaining 11 or 10 or whatever...

BE POSITIVE, THINK POSITIVE, ACT POSITIVE and then EXECUTE!..it WILL be.'

CSA PATTERN RECOGNITION

CATEGORY: ACUTE ADMISSION

Familiarise yourself with malaria, swine flu, Weil's disease, pulmonary embolus, spontaneous and tension pneumothorax, acute and chronic leg ischaemia, acute exacerbation of asthma, acute myocardial infarction, acute subarachnoid haemorrhage, in fact anything acute that you can think of that would warrant a blue light ambulance admission. Remember to offer to call the patient's partner to meet the patient at the hospital.

CATEGORY: CANCER/PALLIATIVE CARE

When doing a role-play home visit on a CSA actor with metastatic cancer, please ensure you do not hover over him but rather crouch down or kneel by his bed, if no chair is present.

Remember that euthanasia is illegal so do not be complicit in prescribing excessive painkillers for a palliative care patient.

CATEGORY: CARDIOVASCULAR

If in doubt, remember you may say that you intend to take advice from the cardiology registrar on call.

If a CSA actor refuses to go to hospital, remember to ask '*why?*' The Medical Protection Society says that GPs do not ask '*why*' enough. This may indicate that the case is not a cardiovascular case but one testing capacity to refuse to go to hospital, ie to refuse medical treatment.

If you wish to commence medication and the patient refuses, you could also try to negotiate a trial use of the drug for a month. The patient may agree to a compromise.

It is a fine balance between reassuring a patient with words or with investigations and appearing to waste NHS resources. The rule of thumb is that if you have failed to reassure the CSA actor patient and he is still very anxious, then you may have to agree to an investigation to give the actor peace of mind.

CATEGORY: DERMATOLOGY

Familiarise yourself with urticaria.

Familiarise yourself with the indications of use of an epipen.

CATEGORY: DIABETES

Familiarise yourself with the exemptions from fasting for Ramadan and also the latest DVLA guidance on HGVs and IDDM.

Be able to use a monofilament to test sensation in a diabetic lower limb.

CATEGORY: DOMESTIC VIOLENCE/RAPE

Domestic violence is very prevalent and many GP surgeries are aware that they are not detecting this as well as they could. A CSA actor may portray herself as a victim of domestic violence or rape by avoiding eye contact, appearing withdrawn or even weeping. Do observe for any nonverbal cues.

It is vital to impart that domestic violence is illegal and that victims may call the police. Offer leaflets on the National Domestic Violence 24-hour helpline number, local women's shelters and victims support.

CATEGORY: ELDERLY

Inform elderly patients of the home visit policy of calling the surgery first to discuss need of a home visit. Sometimes in reality, elderly patients make appointments to see the GP simply because they are lonely. Suggest Age UK to patients. They organize social activities. The elderly are also often carers for their ageing partners. Remember to consider respite care if they feel overwhelmed.

The elderly often have a limited pension. It may be worth checking the diet of an elderly patient to ensure they are not anaemic from poor diet.

CATEGORY: EQUALITY

The NHS has a gatekeeper role. Some cases may challenge you to stick to NHS rules. Circumcisions for religious reasons are not available on the NHS. You may suggest a Muslim speak to his local mosque to find a Muslim GP who performs religious circumcisions. It may be a good idea to ensure both parents consent for their child although the mother is the legal guardian.

CATEGORY: ETHICS

Some cases may ask you to delete information from practice records. Discuss practice staff confidentiality and probity; it is illegal or against the ethical code to delete or tamper with information in practice records. Some practices do offer to put sensitive information in a 'hidden' folder. Read coding also offers another form of anonymity.

A family member may wish to arrange for a nursing home for an elderly parent. Remember to check the capacity of the elderly parent. He may make his own decisions if he is able to understand the facts presented and retain the information, and he is seen being able to weigh up the pros and cons in a decision-making manner. Respect patient autonomy.

You may be asked to prescribe an unlicensed herbal remedy or unlicensed homeopathic treatment. Remember my Golden Rule of CSA and politely explain the licensed treatments and test capacity of the actor patient to refuse conventional medical treatment. If the treatment is available over the counter, this is another option to offer.

CATEGORY: ENT

Be able to use a tuning fork to diagnose a unilateral hearing loss.

Be able to perform a cerebellar exam to be able to diagnose viral labyrinthitis and ensure the patient has someone to drive them home from surgery. You may need to offer an antiemetic in surgery also.

For patients who regularly swim and have recurrent ear infections, it may be a good idea to suggest custom-made ear plugs.

CATEGORY: EYES

Be able to diagnose glaucoma, papilloedema and other common eye conditions from a fundoscopic image on the ipad.

Be able to perform a visual field check to make a diagnosis.

CATEGORY: GENETICS (see last page)

CATEGORY: HEALTH PROMOTION/EDUCATION

You may be presented with cases that simply ask you to educate a CSA actor on diabetes, high blood pressure, epilepsy, the pill, hormone replacement therapy, PSA testing, HIV testing, eczema, psoriasis, etc. The reported age of the CSA actor gives you a clue. If the stated age is in their 20s, it implies the actor will play a patient with poor understanding or even misunderstanding of his or her condition, and it is for you to correct and improve their understanding.

A rule of thumb is that if a CSA actor appears with a chronic condition, the underlying challenge is often one of detecting non-compliance with medication and a poor understanding of the condition.

When asking about alcohol, do not assume what the national limits of a particular country are, ask for quantity. Also be aware that binge drinking may also be a presentation of alcohol addiction.

CATEGORY: LEARNING DISABILITIES

The CSA exam does not employ children at this time, so instead they may have parents come as a third party consultation or have CSA actors role-play as a patient with a learning disability. Often times CSA may wish you to check the capacity of the adult with learning disability, as they may have capacity to refuse treatment. Do not be tempted to hide medication in food. Always check the capacity of the patient as he or she may have capacity, even if schizophrenic (case precedent: Re C case).

CATEGORY: MOTHER'S CONCERN

A common underlying concern of a mother is that her child will be taken away if she is unable to physically or financially support her family. Be prepared to reassure the mother that this will not be the case. Try to reassure mother that even in cases of domestic violence or drug misusers, child protection services will work with the family to try to support the family and keep the family together.

CATEGORY: NORMAL EXAM

Many times patients present with a multitude of symptoms and yet on physical exam there is nothing to see. Be open to psychological causes of symptoms. Stress may play a role as well as recent bereavement and depression. If the examination card reads normal, then think psychological causes.

CATEGORY: PATIENT SAFETY

Some cases may involve a patient presenting with a long list of complaints. Be ready to hear what is on the entire list, so that you do not miss the red flag!

Sudden death questions. This I call questions imposed by the CSA actor in which giving the wrong answer may result in a clear fail or sudden death jeopardy! Be very wary of these so called 'sudden death' questions. A rule of thumb is that we cannot forbid a patient from travelling when ill or going to a disco with strobe lights if they have epilepsy. Here explain the medical risks, so that the CSA actor patient may make a fully informed decision whether it is wise to attend a family event rather than go straight to casualty.

If a patient presents symptomatic with no obvious cause, ensure you ask to review any drug lists for potential side effects, overdose or toxicity.

CATEGORY: PAEDIATRICS

At present there are no child actors in CSA. A parent may come on behalf of a child as a third party consultation. Do ask for the name of the child and his or her age. If the parent cannot bring in a child for an examination and the parent is adamant he or she would like medication, please be flexible and prescribe if the history fits the diagnosis. Stubborn candidates find themselves failing this category for being too rigid.

Do familiarize yourself with paediatric conditions such as ADHD, Asperger's, bedwetting, child abuse, etc. Remember that any suspicion of child, vulnerable adult or elder abuse must be reported to Child Protection Services or Social Care, respectively.

CATEGORY: RHEUMATOLOGY and MUSCULOSKELETAL

Often times the quickest way a self-employed person with a shoulder or elbow tendonitis or hip bursitis may get back to work is by offering a steroid injection.

CATEGORY: SEXUAL HEALTH

Familiarise yourself with the common sexually transmitted diseases, chlamydia, herpes, etc. Be able to explain what these conditions are. Try to encourage GUM clinic (men's sexual health clinic) for contact tracing and a full STD check up. Impress upon the CSA actor that all tests are confidential and the consultation itself is confidential.

CATEGORY: SICK NOTES

You may be asked for an inappropriate sick note. Remember you may issue a sick note for a short period for 'stress at work' or 'stress at home'. Determine what the stressors are and how it is affecting the CSA actor psychologically.

Patients may ask work to give them special compassionate leave to attend a funeral so no sick note is required. If the patient is not present, then you may offer to write a factual letter for absence. Bear in mind this is non-NHS work and attracts a fee.

CATEGORY: TEENAGERS

Be ready to cope with bullying, under age sex, deliberate self harm, and drugs. Do ask the age of the boyfriend and be aware that a 18-year-old having sex with a minor (under the age of 16) is committing a Schedule 1 sex offence and goes on the sex offender's register. Anyone found having sex with a child 12 or under is committing statutory rape with a mandatory jail sentence.

CATEGORY: TELEPHONE

Often times, a patient will ring and their underlying concern requires a home visit for complete reassurance. Consider also bringing the patient to the surgery or out of hours centre if the patient is able to travel.

Be aware of the time scale in which a child should be seen depending on what signs and symptoms they have.

It is reasonable to suggest Calpol to the mother if the child has no red flag signs and the mother is happy to try this first, but do ensure you safety net with a call back and with instructions for mother to ring you if the situation changes.

In most cases of telephone consultations, it will end with a home visit or a visit to the location of the patient.

Be careful not to breach confidentiality over the phone. Always ensure you have the patient's consent to disclose information to third parties.

As the examiner and actor sit in another room, ensure you have a soft, gentle tone of voice and speak slowly and clearly into the phone.

CATEGORY: TWO-WEEK REFERRAL

Asking the red flags is included in my 10 CSA Commandments so make it a habit to ask even if the CSA actor patient asks for something trivial. We have to be sure we are not missing something more worrying.

Remember the age of a patient may in itself be a red flag when a condition presents for the first time.

CATEGORY: VALUING DIVERSITY

You may be seeing CSA actors who role-play as deaf or blind patients. Do not utter 'oh you're BLIND' as you open the door. Offer assistance to a blind person. Do not shout or raise your voice at a 'deaf' CSA actor. Signal whether they lip read. You may end up writing out your questions and the actor responding back in writing if he or she does not communicate by lip reading! Paraphrase any written questions to save on time.

PALLIATIVE CARE MEDICATION
(Source: Calderdale and Huddersfield NHS Trust:
The Syringe Driver in Palliative Care)

ANALGESICS

Diamorphine is the opioid of choice but may not be available, in which case morphine sulphate should be prescribed.

Initial starting dose is dependent on existing opioid requirements. If opioid-naïve, give diamorphine or morphine sulphate 2.5-5 mg prn subcutaneously - if patient requires > 3 prn doses over next 24 hrs, consider a syringe driver. The starting dose should not exceed 10 mg/24 hrs.

If already on oral morphine sulphate, divide the total daily dose by 3 to convert to diamorphine:
> e.g. Zomorph 60 mg bd = 120 mg daily ≡ diamorphine 40 mg/24 hrs via syringe driver.

Ensure that adequate breakthrough doses are written up prn - they should be the equivalent of 1/6 of the 24 hr driver dose (i.e. 7.5 - 10 mg diamorphine prn in above example).

If diamorphine is unavailable, divide the total daily dose by 2 to convert to subcutaneous morphine sulphate:
> e.g. Zomorph 60 mg bd = 120 mg ≡ morphine 60 mg/24 hours via syringe driver.

Ensure that adequate breakthrough doses are written up prn - they should be the equivalent of 1/6 of the 24 hour driver dose (i.e. 10 mg morphine prn in above example).

To convert from other opioids, contact pharmacy or the Palliative Care Team.
(Sample prescription:

Diamorphine injection 30mg/ml ampoule, size 1 ml.
40mg subcutaneous infusion over 24 hour in syringe driver.
21 (TWENTY ONE) ampoules.

Water for injection 10 ml ampoules x 10,
Via syringe driver over 24 hours, to be supplied as needed.)

ANTI-EMETICS

Different anti-emetics have different sites of action. Two of the most commonly used anti-emetics are cyclizine and haloperidol. Cyclizine is a broad spectrum anti-emetic useful when the cause of nausea is unclear. Haloperidol is the anti-emetic of choice for opioid induced nausea, or nausea associated with uraemia or hypercalcaemia.

> CYCLIZINE available in 50 mg ampoules
> Dose range 100-150 mg/24 hrs
> Compatible with diamorphine
> May precipitate if concentration of cyclizine or
> diamorphine > 25 mg/ml

> HALOPERIDOL available in 5 mg ampoules
> Dose range 5 -10 mg/24 hrs
> Compatible with diamorphine

If symptoms persist after 24 hrs, contact pharmacy or Palliative Care Team - alternative anti-emetics may be indicated.

SEDATIVES

Benzodiazepines are extremely useful for the management of terminal agitation or restlessness. Midazolam has a very short duration of action and for sustained effect requires to be given in a continuous infusion.

MIDAZOLAM available as 10 mg/2 ml ampoules
> Starting dose 10 mg/24 hrs.
> Increase in 10 mg/24 hr increments to maximum 60 mg/24 hrs. Prescribe prn dose 2.5 mg - 5 mg 2 hourly

ANTIMUSCARINICS

These drugs are useful for the relief of respiratory tract secretions and for the relief of intestinal colic.

> HYOSCINE BUTYLBROMIDE (Buscopan®)
> Available as 20 mg/ml ampoules
> Starting dose 30-60 mg/24 hrs.
> Max dose usually 120   mg/24 hrs.
> Prescribe prn dose 20 mg 4 hrly prn.

Statement of Fitness for Work
For social security or Statutory Sick Pay

Patient's name	Mr, Mrs, Miss, Ms
I assessed your case on:	/ /
and, because of the following condition(s):	

I advise you that:

☐ you are not fit for work.

☐ you may be fit for work taking account of the following advice:

If available, and with your employer's agreement, you may benefit from:

☐ a phased return to work   ☐ amended duties

☐ altered hours   ☐ workplace adaptations

Comments, including functional effects of your condition(s):

Sample

This will be the case for	
or from	/ / to / /

I will/will not need to assess your fitness for work again at the end of this period.
(Please delete as applicable)

Doctor's signature	
Date of statement	/ /
Doctor's address	

Med 3 04/10

5
6

Sick Notes / Medical Certificates: Info for patients and their employers
Do you need a doctor's certificate in the first seven days of an illness?
No. Illnesses which last less than one week are often minor and self limiting and you may not require a visit to a doctor. It can be difficult for your doctor to judge whether or not you are incapable of reporting for work in this situation and all the certificate really indicates is that you attended the surgery on a specific date complaining that you had an illness. In general, the Departments of Health and Social Security, employers, doctors, and patients do not recommend that you attend your doctor for sickness certification alone. **Your GP is only required to issue a certificate if your absence from work through sickness lasts more than seven days.**
What do you need?
For the **FIRST FOUR DAYS** of any illness you do not require any form of certification. (It is for you to decide if you are fit to work).
For the **NEXT THREE DAYS** (including Saturdays and Sundays) you must fill in a self certification form, (**SC2** which is available from your employer *or complete the form below*).
After the **FIRST SEVEN DAYS** you will need a doctor's certificate and will have to attend the doctor to get a Department of Social Security sick note either a 'Med 3' or 'Med 5'.
What are private sickness certificates and do you need one?
Some employers or insurance schemes will ask you to provide a private sickness certificate. Most surgeries **will charge you a fee to** provide you with a private sickness certificate.
DSS Sick notes are not usually necessary for the first week of any illness.

- - - - - - - - - - - - - - - - - -

PRIVATE MEDICAL CERTIFICATE

This is to certify that in the opinion of ……………………………………………
……………………………………………of
………………………………………………………………………………
is / was suffering from…………………………………………………..
and is unable to attend work/school.
Signed…………………………………………

Date……………………………………….   Surgery stamp

Please Note Medical certificates are not required for absences under 7 days and employers should normally accept self certification. Certification for self limiting illness by a doctor wastes NHS time.

CERTIFICATE

NOT VALID
UNLESS
STAMPED

EMPLOYEES STATEMENT OF SICKNESS (SELF CERTIFICATE/ SC2)

Surname……………………………Other
Names………………………………………Title……………………..
National Insurance Number……………………………………………Date of
Birth…………………………….
Clock or Payroll Number……………………..…………………….
About your sickness: Please give brief details of your sickness

What date did your sickness begin…………….……….What date did your
sickness end……………..…………
Was your sickness caused by an accident at work or by an industrial
disease? **yes / no**
If you answered yes you may be able to get Industrial Injuries Disablement
Benefit. If you want information about this benefit, ask at your nearest
Department for Work and Pensions.

Your Signature……………………………………………..Date……………………..

Self Certification for Candidates who have missed an Examination

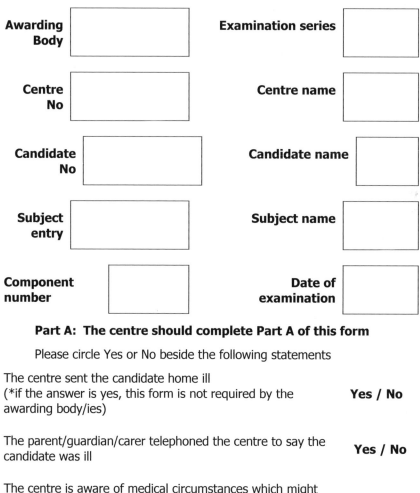

Awarding Body		**Examination series**
Centre No		**Centre name**
Candidate No		**Candidate name**
Subject entry		**Subject name**
Component number		**Date of examination**

Part A: The centre should complete Part A of this form

Please circle Yes or No beside the following statements

The centre sent the candidate home ill
(*if the answer is yes, this form is not required by the awarding body/ies)   **Yes / No**

The parent/guardian/carer telephoned the centre to say the candidate was ill   **Yes / No**

The centre is aware of medical circumstances which might cause absence
(*if the answer is yes, this form is not required by the awarding body/ies)   **Yes / No**

The candidate has missed an examination in a terminal series   **Yes / No**

Head of centre/Exams officer

Date   _____

Name (Please print)

Signature   _____

_____

Part B: The doctor/nurse or surgery receptionist should complete Part B

Please circle Yes or No beside the following statements

The patient was seen in the surgery at reception	**Yes / No**
The patient was seen by the nurse	**Yes / No**
The patient was seen by a doctor	**Yes / No**
The patient did not attend the surgery but the doctor/nurse spoke to the parent/ guardian/carer on the telephone	**Yes / No**
The patient was thought to be unfit to sit examinations	**Yes / No**

Any other relevant information

Signed by member of surgery staff

Date   _____

Name (Please print)

Signature   _____

_____

6
0

Part C: The parent/guardian/carer should complete Part C

Please circle Yes or No beside the following statements

I telephoned the School/College on the day of the examination to
say that my son/daughter/ward was too ill to take an examination

Yes / No

I telephoned the surgery to let them know the symptoms and
receive advice

Yes / No

The symptoms
were: _____

_____

_____

Declaration by parent/guardian/carer
I understand that it is fraudulent to claim that a candidate is ill when he or she is
fit to attend for a scheduled examination.
I understand that the results can be withdrawn and the candidate disqualified if
fraudulent claims are made.

Signed by parent/guardian/carer

Date

_____

Name (Please print) _____

Signature _____

6
1

Part D: The candidate should sign Part D

Declaration by candidate
I felt too ill to attend my examination.
I understand that my results can be withdrawn or I can be disqualified if I claim to be ill when I am not ill.

Signed by candidate

Date _____

Name (Please print)
_____

Signature
_____

6
2

Notes on the Use of the Self Certification Form

This is not a special consideration form. It does not need to be submitted if the centre knows that the candidate is ill. This form is NOT required in the following circumstances:

- the candidate has missed a module test and can re-enter at a later date;

- the candidate was sent home ill by the centre;

- the candidate was seen to be falling ill in the centre the day before the absence;

- the centre knows of long-term medical circumstances which can lead to sudden absence;

- the candidate has a prescription or label from medication showing the date when the medicine was prescribed and the name and address of the candidate;

Where the centre can verify the circumstances, they should be stated on the special consideration form (JCQ/SC – Form 10).

This self certification form should be used only in the following circumstances:

- the candidate has missed a terminal examination or a module which cannot be re-entered;

- the centre has no reason to suspect that this may be a fraudulent claim;

- the candidate has been attending other examinations so far without problems.

Procedure

The form should be kept in the centre and Part A completed only where medical evidence is required and when the parent/guardian/carer telephones the centre and the surgery to say what has taken place.

The candidate/parent/guardian/carer takes the form to the surgery for Part B to be completed.

The parent/guardian/carer completes Part C and the candidate completes part D.

This form does not replace the special consideration form. It should be attached to the special consideration form (JCQ/SC – Form 10) relating to a missed examination in a terminal series.

References:
RCGP Curriculum,
Dr Una Coales's PLAB: 100 OSCEs book
Dr Una Coales's DRCOG: How to Pass First-Time book
NICE guidances/ BMJ/ BJGP peer-reviewed journals

Abbreviations: DG: data gathering, IPS: interpersonal skills, CMx: clinical mx skills; HP = health promotion; DTA (discuss tobacco and EtOH); EdxCU - explain diagnosis and check understanding; FH = family history; FU = follow-up; IADL: impact on activities of daily living; Ix: investigations; PILS = offer patient info leaflet; OE = on exam; PN = practice nurse; RGO = reassure good outcome; SEmp = self-employed; SocS = social support.

CV/ CIRCULATORY SYSTEM

1. Vertebrobasilar or Carotid TIA/ Stroke 70 yo M (NICE 08)

DG: transient dizziness, visual loss, N/V, ataxia vs. hemiparesis, hemiplegia, amaurosis fugax. FH + CVA/MI. Job? Home? R/O CVA. OE: fundi, BP/P, AF, carotid, CNS weakness. IPS: Family concerned. Expectations? Empathy. IADL. EdxCU: 'Mini' stroke vs. CVA. CMx: Espirit and Heart Study: ASA 300mg + dipirydamole 25mg tds x 2/52, simvastatin 40 mg (low cholesterol diet). DVLA ban 4/52, DTA. PILS. FU family. Refer rapid access TIA (blood, CT, ECG, doppler carotid) within 1/52 or to A+E if ongoing.

Lancet Feb 2007; 369:283-92. Test could predict 2-day stroke risk Doctor mag. $ABCD^2$ score. 70,000 diagnosis TIA in UK, up to 20% CVA w/n 90d and ½ w/n 2 days. Used $ABCD^2$ score in 1,916 TIA patients. High risk 5/7 patients and benefit from intervention w/n 24h. Low risk < 4 patients, no hosp observation. $ABCD^2$ score RF: age ≥ 60 (1), SBP ≥ 140 or DBP ≥ 90 at 1st assessment after

TIA (1), unilateral weakness (2 patients), speech impairment without weakness (1), $\geq$ 60 mins TIA (2), 10-59 mins (1), DM (1).

BMJ 28 April 2007: Give dipyridamole with ASA instead of ASA alone to prevent vascular events after ischaemic stroke or TIA. Cathie Sudlow. ESPS-2 (2nd European stroke prevention study) 3299 patients with TIA or prior ischaemic stroke dipyridamole + ASA ↓ RR of vascular events by 1/5th vs. ASA alone but ASA alone was 50 mg vs. 300. ESPRIT (European/ Australasian) 2739 patients showed ↓RR with combo and with meta-analysis. Add dipyridamole prevented 10/1000 patients events. UK cost barriers dipyridamole MR 200 mg bd £102/y vs. ASA 75 mg od £5/y.

BHF Factfile Dec 2006 Early intervention in stroke IV thrombolysis (with tPA) w/n 3h of onset improves outcome in ischaemic stroke. Before tx, CT to r/o haemorrhage and BP should be < 185/110. Once CT r/o bleed, start ASA 300 mg PR or PO and then 75-300 mg od. Early recurrent stroke risk is high. 2°prev important. Admission to specialised stroke unit ↓mortality by 20-25%. Risk of DVT↑ so elasticated stockings and low-dose SC heparin. Swallowing assessment (aspiration pneumonia risk so NGT). Addition of 300 mgs bd dipyridamole to ASA, statin and indamapide and perindopril all↓ recurrent CVA even in normal TC and BP. Carotid endarterectomy within 1-2 weeks of event for maximum benefit or ASAP after TIA and minor stroke.

BMJ 9 Sept 2006: Under investigation and under treatment of carotid disease in elderly patients with TIA and stroke: comparative population based study. JF Fairhead et al. Oxfordshire (Apr '02-Mar '03) and OXVASC pop (Apr '02 to Mar '05). Compared with patients in OXVASC study, rates of carotid imaging, diagnosis of $\geq$ 50% symptomatic stenosis, and carotid endartectomy in $\geq$ 80 yo were substantially ↓. Sx carotid stenosis ↑ steeply $\geq$ 80. Carotid imaging part of TIA ix!

Use sudden onset of neurological sxs a validated tool, FAST (Face Arm Speech Test), to screen for a diagnosis of stroke or TIA. R/O ↓ glucose. Suspected TIA who are at high risk of stroke (ABCD^2 score of ≥ 4) should have: aspirin 300 mg od ASAP, specialist assessment and ix within 24 hours of onset, and measures for 2° prevention as soon as the diagnosis is confirmed. Treat crescendo TIA (≥ 2 TIAs/week) as high risk of stroke, even with an ABCD^2 score of ≤ 3. Some who have had a stroke or TIA, have narrowing of the carotid artery that may require surgery. Carotid imaging is required to define the extent of carotid artery narrowing. Suspected TIA (resolved < 24 hrs) should be assessed by a specialist (<1 week of onset) before a decision on brain imaging is made.

Patients who have had a suspected TIA who are at high risk of stroke (an ABCD^2 score of ≥ 4, or with crescendo TIA) in whom the vascular territory or pathology is uncertain should undergo urgent diffusion-weighted MRI. All with suspected non-disabling stroke or TIA who after specialist assessment are considered as candidates for carotid endarterectomy should have carotid imaging < 1 week of onset of sxs. Patients who present > 1 week after their last sx of TIA has resolved should be managed using the lower-risk pathway.

Refer for urgent carotid endarterectomy and carotid stenting for stable neurological sxs from acute non-disabling stroke or TIA who have symptomatic carotid stenosis of 50–99% according to the NASCET (North American Symptomatic Carotid Endarterectomy Trial) criteria, or 70–99% according to the ECST (European Carotid Surgery Trialists' Collaborative Group) criteria w/n 1 week of onset of stroke or TIA sxs; undergo surgery within a maximum of 2 weeks of onset of stroke or TIA sxs; receive best medical tx (BP control, antiplatelet agents, cholesterol lowering via diet and drugs, lifestyle advice).

All suspected stroke should be admitted directly to a specialist acute stroke unit following initial assessment, from the community or from A&E. MRI ix if: indications for thrombolysis or early anticoagulation tx, on anticoagulant tx, a known bleeding tendency, ↓ LOC (GCS < 13), unexplained progressive or fluctuating sxs, papilloedema, neck stiffness or fever, or severe headache at onset of stroke sxs.

Thrombolysis with alteplase recommended for the tx of acute ischaemic stroke by doctors trained and experienced in the mx of acute stroke. Aspirin and anticoagulant tx - with acute ischaemic stroke. All who have had a diagnosis of 1° intracerebral haemorrhage excluded by MRI should be given, ASAP or < 24 hrs: aspirin 300 mg po if they are not dysphagic or aspirin 300 mg PR or by enteral tube if they are dysphagic. Thereafter, continue aspirin 300 mg until 2/52 after the onset of stroke, and then initiate definitive long-term antithrombotic tx. Patients being discharged < 2 weeks can be started on long-term tx earlier. Give PPI if previous dyspepsia associated with aspirin. Anticoagulation tx should not be used routinely for the tx of acute stroke.

2. Chest Pain ➔ left arm 65 yo M

DG: costochondritis vs. angina? FH + IHD. DTA. IPS: Home, job, hobbies, sports. OE: BP, CV, ECG. EdxCU: angina. CMx: Options: aspirin 75 mg, GTN spray, advise if pain lasts > 10-mins not relieved by 2 puffs to call 999, refer rapid access CP clinical for exercise ECG, cxr, bloods (fbc, lipids, u/e, fbg), echo, metoprolol, statin, stress mx, exercise, change job, British Heart Association (self-help), DVLA cease till sxs controlled. Need not notify. Send in MI. Need trop t levels to confirm if < 48 hours.

BMJ 7 July 2007 Advising patients on dealing with acute chest pain A Khavandi et al. BHF GTN spray 3x at 5 min intervals before call 999. Chest pain > 15 mins probably MI. STEMI (sxs of ST elevation MI) 1 GTN spray (2 metered 800 ug) and 5 mins call.

Successful defib ↓7-10% each min after arrest. Median 30 mins before v fib.

Pulse 7 Dec 2006 Need to know Angina A Cormack cardiologist. A negative exercise test is reliable indicator of patent coronary arteries. No evidence for/ against ↓ homocysteine with folic acid. Homocyst RF for CHD. Renal disease, metoprolol preferred β blocker. Carvedilol, bisoprolol and nebivolol if concomitant heart failure. Nitrates have no prognostic benefit.TC < 4 and LDL 2 targets. Minority of angina at rest, after meals or emotional vs. exertional. Echo only if murmur or sxs ht F. Benefit anti-platelets, statins, BP control. Ca channel blocker in β-blocker resistance. ASCOT and NICE: Amlodipine, adalat LA or felodipine for ↑BP. Nicorandil 3rd line. Simvastatin 40 mg may ↑. CABG↓, multi-vessel PCI↑. Metal stents restenosis rate 10%↓with introduction of drug-eluting stents.

GP Mag 29 Sept 2006. Mx of angina L Newson 2 mill UK. 30% suffer MI, death or revascularisation within 2 years. B-blocker 1st line. Ivabradine is 1st selective sinus node I_f inhibitor (Eur Heart J 2005) as effective as atenolol for stable angina. (BMJ 2005- ECG and exercise tolerance test less specific and less sensitive in F vs. M). IONA trial-add nicorandil ↓ CVD incidence (Heart 2006). SIGN: all patients with recent-onset angina refer to cardiologist. NSF CHD. ↑ACEI use.

NICE: MI: 2° PREVENTION IN 1° AND 2° CARE - May 07

Advise regular physically active for 20–30 mins a day to the point of slight breathlessness. If not achieving this, advise to ↑ their activity in a gradual, step-by-step way, duration and intensity of activity. Advise quit and offer smoking cessation assistance. Advise NOT to take supplements containing beta-carotene, or to take antioxidant supplements (vitamin E and/or C) or folic acid to ↓ CV risk. Eat a minimum 7 g of omega 3 fatty acids per week from 2 - 4 portions of oily fish. MI within 3 months and not achieving 7 g of omega 3 fatty acids/week, consider providing at least 1 g daily of omega-3-acid ethyl esters tx licensed for 2° prevention post MI for up to 4 yrs.

Advise to eat a <u>Mediterranean-style diet</u> (bread, fruit, vegetables and fish; less meat; and replace butter and cheese with products based on vegetable and plant oils). Offer advice to all patients who are overweight. < 21U/week M or 14U F, avoid binge drinking (> 3 alcoholic drinks in 1–2 hrs).

<u>Cardiac rehabilitation</u> should be equally accessible and relevant to all patients after an MI, minority ethnic groups, older, lower socioeconomic groups, F, from rural communities and patients with mental and physical health comorbidities. Include health ed and stress mx.

<u>A home based programme</u> validated for patients who have had an MI (i.e. 'The Edinburgh heart manual'; www.cardiacrehabilitation.org.uk/heart_manual/heartmanual.htm) incorporates education, exercise and stress mx components with FUs by a trained facilitator may be used to provide comprehensive cardiac rehabilitation.

<u>Most patients who have had an MI can return to work</u>. Take into account the phys and psychological status of the pt, the nature of the work and the work environment (www.dvla.gov.uk). After an MI without complications, <u>patients can usually travel by air within 2–3 weeks</u>. Patients who have had a complicated MI need expert individual advice.

<u>Patients should be reassured that after recovery from an MI, sex</u> presents no greater risk of triggering a subsequent MI than if they had never had an MI. Patients with an uncomplicated recovery may have sex after 4 weeks. When treating erectile dysfunction, tx with a PDE5 (phosphodiesterase type 5) inhibitor may be considered in patients who had an MI > 6 months earlier and who are now stable.

PDE5 inhibitors must be avoided in patients on nitrates and/or nicorandil - can lead ↓↓BP.

Drug tx: Offer all patients who have had an acute MI a combo of: ACEI, aspirin, b-blocker, statin. Early after MI, offer all an ACEI and titrate up q 1 to 2 weeks until reach the max tolerated or target dose.

Assess LV function in all who have had an MI. After an MI, all with preserved LV function or with LV systolic dysfunction should continue tx with an ACEI indefinitely, whether or not they have sxs of heart failure. Routine rx of ARBs after an acute MI is NOT recommended. Offer an ARB for those intolerant to ACEI due to cough or allergy. U/e, serum lytes and BP should be measured before starting an ACEI or ARB and again w/n 1 or 2 weeks of starting tx. Monitor as the dose is titrated up and then at least annually. Monitor CHF as per NICE.

Antiplatelet therapy: Aspirin should be offered to all patients after an MI, and should be continued indefinitely. Clopidogrel should NOT be offered as 1st -line monotherapy after an MI. Clopidogrel, in combination with low-dose aspirin, is recommended for use in the mx of non-ST-segment-elevation acute coronary syndrome in patients who are at moderate to high risk of MI or death. Monitor u/e and K before and during tx with an aldosterone antagonist. If hyperkalaemia is a problem, halve the dose of the aldosterone antagonist or stop.

After an MI, offer all statin ASAP. Measure baseline lfts before initiation. Patients who develop muscle sxs (pain, tenderness or weakness) - measure creatine kinase. Statins should be discontinued in patients who develop peripheral neuropathy that may be attributable to the statin tx, and further advice from a specialist should be sought. After an ST-segment-elevation MI, patients treated with a combination of aspirin and clopidogrel during the 1st 24 hrs after the MI should continue this tx for at least

4/52. Thereafter, standard tx including low-dose aspirin should be given. For patients with aspirin hypersensitivity, clopidogrel monotherapy should be considered as an alternative tx. In patients with a history of dyspepsia, treat with a PPI and low-dose aspirin. After appropriate tx, patients with a h/o aspirin-induced ulcer bleeding whose ulcers have healed and who are negative for *Helicobacter pylori* should be considered for tx with a full-dose PPI and low-dose aspirin.

Early after an acute MI, all patients without LV systolic dysfunction or with LV systolic dysfunction (symptomatic or asymptomatic) should be offered tx with a b-blocker. B-blockers should be initiated ASAP when the pt is clinically stable and titrated upwards to the max tolerated dose. B-blockers should be continued indefinitely after an acute MI.

For patients who have had an MI, high-intensity (vitamin K antagonist) warfarin (INR >3) should NOT be considered as an alternative to aspirin in 1st-line tx. If unable to tolerate either aspirin or clopidogrel, tx with moderate-intensity warfarin (INR 2–3) should be considered for up to 4 yrs, and possibly longer. For patients who have had an acute MI, are intolerant to clopidogrel and have a low risk of bleeding, tx with aspirin and moderate-intensity warfarin (INR 2–3) combined should be considered. For patients already being treated for mechanical valve, recurrent DVT, atrial fib, or LV thrombus, warfarin should be continued. For patients treated with moderate-intensity warfarin (INR 2–3) and who are at low risk of bleeding, add aspirin. The combination of warfarin and clopidogrel is NOT routinely recommended.

Ca channel blockers should NOT routinely be used to ↓ CV risk after an MI. If b-blockers are C/I or need to be discontinued, diltiazem or verapamil may be considered for 2° prevention in patients without pulmonary congestion or LV systolic dysfunction. For patients who are stable after an MI, ca channel blockers may be used to treat hypertension +/or angina. For patients with heart failure, use amlodipine and avoid verapamil, diltiazem and short-acting dihydropyridine agents in line with CHF (NICE CG 5).

Nicorandil (K channel activator) is NOT recommended to ↓ CV risk in patients after an MI. Acute MI and who have sxs and/or signs of heart failure and LV systolic dysfunction, tx with an aldosterone antagonist licensed for post-MI tx should be initiated w/n 3–14 days of the MI, preferably after ACEI. Patients who have recently had an acute MI and have clinical heart failure and LV systolic dysfunction, but who are already being treated with an aldosterone antagonist for a concomitant condition (i.e., chronic heart failure), should continue with aldosterone antagonist or alternative, licensed for early post-MI tx. For patients who have had a proven MI in the past and heart failure due to LV systolic dysfunction, tx with an aldosterone antagonist should be in line with (NICE CG 5).

All patients should be offered a <u>cardiological assessment to consider whether coronary revascularisation</u> is appropriate. This should take into account comorbidity. <u>Hypertension should be treated</u> to the currently recommended target of ≤ 140/90 mmHg. Patients with relevant comorbidities, i.e. diabetes or renal disease, should be treated to a lower BP target. Patients who have LV systolic dysfunction should be considered for an implantable cardioverter defibrillator in line with 'Implantable cardioverter defibrillators for arrhythmias' (NICE).

<u>BMJ 5 May 2007 B blockers in HTN and CV disease</u> HT Ong. ASCOT-BPLA (BP lowering arm) atenolol marginally inferior to amlodipine for HTN, 19 257 patients amlodipine + perindopril vs. atenolol + bendrofluazide. Amlodipine had ↓ stroke, mortality, coronary end point, glucose, creatinine, TG, BP; higher HDL. CIBIS-II for heart failure bisoprolol vs. placebo ↓ hospitalisation and death from CV causes. MERIT-HF, metoprolol ↓ mortality (CV), sudden death, and death from ht failure. COPERNICUS (2289 patients, NYHA IV) carvedilol ↓ mortality. β blockers

↓mortality after MI and improve prognosis in patients with systolic heart failure.

Lancet 2005; 366:1640-1649. Obesity and the risk of MI in 27,000 participants from 52 countries: a case control study. Followed the INTERHEART study. Waist-to-hip ratio shows a graded and highly significant associated with MI risk worldwide that is superior to both BMI and waist circumference

BMJ 2006:878-882. Clinical value of the metabolic syndrome for long-term prediction of total and CV mortality: a prospective, pop based cohort study. Findings-metabolic syndrome little value in predicting heart disease, more useful predictor of diabetes. HDL not included, which ↓ predictive accuracy of Framingham scoring.

3. Palpitations (NICE ATRIAL FIBRILLATION JUNE 2006)

DG: 65 yo M feeling dizzy, diaphoretic, N/V. Taking ASA + GTN for angina. IPS: Job? Home? Caffeine? Stress? OE: BP, CV, interpret ECG EdxCU: atrial fib vs. ventricular tachyarrhythmia. CMx: NICE 2006 guidance: rhythm / paroxysmal (young) vs. rate controlled/ permanent AF 65 yo (b-blocker, dig, warfarin INR 2.5). Stratify stroke risk. Ix: 24h ECG, echo, refer to cardiologist vs. send in? Unstable AF (Permanent= electrical cardioversion vs. pharma b-blockers or ca antagonists/ life-threatening = emergency cardioversion).

BJGP Sept 2006: The safety and adequacy of antithrombotic tx for atrial fib: a regional cohort study. C Burton et al. 601 patients. 27 GPs in Scotland. Adequacy of anticoagulation control was broadly comparable to that in clinical trials but risk of severe bleeding was higher, reflecting older age and comorbidities of cohort. 264 died w/n 5 yrs. Annual risk of bleeding on warfarin was 9% similar to 91.2% RCTs vs. severe bleed 2.6% vs. 1.3%.

4. NICE: Mx OF HTN IN ADULTS IN 1° CARE Aug 2004

>140/90 - reassess BP from 2 further visits. Target BP is 140/90. Offer drug tx if ≥160/100 OR Persistent BP > 140/90 + 10 yr CHD risk ≥ 15% OR CVD ≥ 20% OR existing CV disease or target organ. Immediately refer accelerated (malignant) HTN and suspected pheochromocytoma.

Offer patients > 80 yo and patients with isolated SBP, same tx as patients with both ↑ SBP and DBP. Lifestyle advice - caffeine, alcohol, diet, exercise, low dietary sodium, stop smoking.
Ix - urine for protein, blood (glucose, lytes, creatinine, TC, HDL cholesterol), 12-lead ECG.
Refer <30 yo, worsens suddenly, BP > 180/110 + signs of papilloedema +/or retinal haem or responds poorly to tx.↑ creatinine (renal disease). Labile or postural hypotension, HA, palpitations, pallor, diaphoresis (pheo).↓K, abdo or flank bruits or ↑ creatinine when starting ACEI suggests renovascular hypertension. Cushing syndrome - osteoporosis, truncal obesity, moon face, purple striae, muscle weakness, easy bruising, hirsutism, ↓K and hyperlipidaemia.

β-blockers may be considered in younger patients: those with an intolerance or CI to ACEI and AIIR antagonists, or women of child-bearing potential or patients with evidence of ↑ sympathetic drive.

In patients whose BP is well-controlled (≤ 140/90) with a β-blocker, no need to replace the β-blocker with alternative.

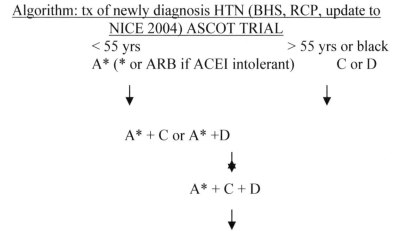

Algorithm: tx of newly diagnosis HTN (BHS, RCP, update to NICE 2004) ASCOT TRIAL

< 55 yrs	> 55 yrs or black
A* (* or ARB if ACEI intolerant)	C or D

A* + C or A* +D

A* + C + D

Add: further diuretic rx or selective α blocker or β blocker. Consider seek specialist advice.

5. Chronic Heart Failure (NICE July 2003)

DG: Hx, sxs (breathlessness, fatigue), signs (fluid retention).
Ix: 12-lead ECG +/or BNP or NTproBNP.
Other recommended tests: cxr, u/e, creatinine, FBC, TFTs, LFTs, FBG, lipids, UA, PFR or spirometry.
If ECG or BNP is abnormal, arrange echo to r/o valve disease, assess sys + diastolic function of the LV and detect intracardiac shunts. If echo normal, consider diastolic dysfunction, refer. If abnormal echo, assess severity, aetiology, ppt factors, type of cardiac dysfunction, consider refer. CMx: Aerobic exercise training/ Refer to smoking cessation services/ Abstain from EtOH if EtOH related. Annual flu jab; Pneumococcal vaccine x 1/ Air travel depends on clinical condition at time of travel/ DVLA.

LV systolic dysfunction Generalist: Start ACEI and titrate upwards q 2/52. Check urea, creatinine and lytes after initiation and at each dose increment. Or if not tolerated (severe cough), consider AIIR antagonists. Add diuretic to control congestive sxs and fluid retention. Diuretic is 1st line if patient presents with acute pulmonary oedema.

Specialist: Add β-blocker and titrate upwards after diuretic + ACEI. Start low, slowly, assess HR, BP and clinical status after each titration. Add digoxin if patient in sinus rhythm remains symptomatic despite tx with diuretic, ACEI + β-blocker OR if patient is in atrial fib then use as 1st-line. Avoid triple tx of ACEI, β-blocker + AIIR antagonist pending results of further trials. Add spironolactone (12.5-50mg od) if pt remains mod to severely symptomatic despite optimal drug tx. Check K, creatinine for signs of ↑K (1/2 dose of spironolactone and recheck) and renal failure.

Amiodarone - by specialist. Need q 6/12 clinical review, LFTs, TFTs, S/Es. Anticoagulants - heart failure + atrial fib; heart failure + h/o thromboembolism, LV aneurysm or intracardiac thrombus. Aspirin (75-150 mg od) for patients with heart failure + atherosclerotic art disease (including coronary heart disease).

Statins for patients with heart failure + atherosclerotic vascular dis. Isosorbide/ hydralazine combo who are intolerant of ACEI or ATIIR antagonist (specialist only). IV inotropic agents (dobutamine, milrinone or enoximone) short-term tx for acute decompensation of chronic heart failure.
Ca channel blockers - amlodipine for tx of co-morbid HTN +/or angina in patients with ht F. Avoid verapamil, diltiazam or short-acting dihydropyridine agents.

Coronary revascularisation - refractory angina. Cardiac transplantation - severe refractory sxs or refractory cardiogenic shock. Cardiac resynchronisation tx – LVEF ≤ 35%, drug refractory sxs, and a QRS duration > 120 ms. Implantable cardioverter-defibrillators (ICDs) - cardiac arrest due to VT or VF, spontaneous sustained VT causing syncope, sustained VT without syncope/ cardiac arrest and who have EF < 35% but are no worse than Class III NYHA; h/o prior MI + non-sustained VT on Holter, inducible VT on electrophysiological testing, LVEF < 35% and no worse than Class III; familial cardiac condition with high risk of sudden death (long QT syndrome, hypertrophic cardiomyopathy, Brugada syndrome, arrhythmogenic RV dysplasia and following repair of Tetratology of Fallot)

Clinical review q 6mos: Hx, NYHA class/ PE - weight, JV distension, lung crackles, HPM, peripheral oedema, lying + standing BP; 12-lead ECG or 24h Holter if suspicion of arrhythmia/ Blood tests - urea, lytes, creatinine. TFTs, FBC, LFT and level of anticoagulation may be required depending on rx. No routine digoxin monitoring. A digoxin concentration w/n 8-12h of last dose to confirm clinical impression of toxicity or non-compliance.

Referral for specialist advice: Heart failure due to valve disease (no ACEI), diastolic dysfunction (Diagnosis + tx by specialist. Treat with low to med dose loop diuretics < 80 mg frusemide /day. If not respond, refer), or any other cause except LV systolic dysfunction. ≥1 co-morbidities (COAD/asthma/ rev airways disease; creatinine > 200, anaemia, thyroid disease, PVD, urinary frequency, gout).

Angina, atrial fib (Specialist to decide whether to improve HR control or cardiovert (return to sinus). Anticoagulation is indicated), or other symptomatic arrhythmia. Severe heart failure/ heart failure not responding to tx as discussed in guideline/ Ht F that cannot be mx in the home. Women who are planning a pregnancy or who are pregnant.

Doctor Mag 10 October 2006 How useful are natriuretic peptide tests in selecting ht F patients for referral? BHF 676,500 ≥ 45 have heart failure. 297 patients in BJGP 2006 FuatA et al The diagnosis accuracy and utility of BNP test in community population of patients with suspected heart failure. NTproBNP performed marginally better than BNP and easier to use in 1°care.

BJGP May 2006;526:327-33. A Fuat et al. The diagnostic accuracy and utility of a B-type natriuretic peptide test in a community population of patients with suspected heart failure. Usefulness of BNP and N-terminal BNP (NTproBNP) in 114 patients with LV systolic dysfunction of 297 patients. NPV for NTproBNP was 92% and for BNP was 88%. 25% ↓ referrals to clinic/ echo.

Pulse 27 April 2006 Dr Kirin Patel. Need to know heart failure. Persistent AF leads to structural and electrical remodelling so DCCV should be attempted early if a patient meets criteria. Up to 15% may be ACEI intolerant. A normal BNP makes heart failure very unlikely but a ↑BNP may be due to other causes, triage for echo.

6. BMJ 27 May 06 Mx of hypertrophic cardiomyopathy P Spirito et al. Most common genetic heart disease. Marked asymmetric LVH, non-dilated LV cavity, diastolic impairment. Commonest cause of sudden cardiac death in young and athletes. Cardioverter defibrillator only effective tx for prevention of sudden D. Medical tx-B-blocker or verapamil improves sxs of heart failure but does not modify clinical course. 5% unresponsive to med tx = surgical myectomy or alcohol septal ablation to relieve outflow obstruction. PAF or chronic AF in minority.

7. BMJ 2006;332:144. The use of statins and other meds for 2^0 prevention of CHD has ↑ since CHD NSF.

NEJM 06;354:1,264-72.↓LDL by 15%-1/2s CHD risk. 15 yrs. 3,363 black+9,524 Caucasian. JBS < 2 LDL.

BHF JBS2 guidelines Jan 06. Tobacco cessation, waist< 102cm M, < 88 cm F, BMI< 25, BP < 130.80, TC < 4, LDL < 2, FBG< 6, HbA1c< 6.5. JAMA 2006;295:180. Patients with ↑ risk CHD poorly tx with ASA, statins and anti-hypertensive meds. JAMA 2006; 295: E1-10. Statins reverse atherosclerosis. ASTEROID study - 40 mgs rosuvastatin /d for 2 yrs. 349 patients. ↓ LDL from 3.3 to 1.6 = 9.1%↓ in vol of coronary atherosclerosis.

8. Intermittent claudication

DG: 72 yo M pain in calves, walks short distances, relieved at rest, smokes cigs x 40 years. OE: Buerger's angle 20° indicates severe ischaemia, hang leg off side to assess capillary filling time, guttering of veins, trophic changes, pressure necrosis, ulcers, temp, peripheral pulse, bruits, assess AAA, BP, ankle-brachial Doppler index < 0.4=severe ischaemia. CMx: send in to hospital urgently if red flags, else refer vascular opinion, mx BP +ASA.

BMJ 7 April 2007 Intermittent claudication RW Simon et al, Zurich. ABPI< 0.9. Tx ASA 75mg od + statins regardless of TC. Weight, tobacco, brisk walk 30 min bd. 5P's pain, pale, pulseless, paraesthesia, paralysis acute limb ischaemia.

Doctor Mag 12 Sep 2006 How should claudication be managed. D Morris. Calf pain relieved by rest. Hang leg out of bed at night (↑blood flow) ABPI< 0.4 critical, > 1 normal. Colour duplex U/S scan locate site and degree of arterial stenosis. 2° prevention: BP, TC, ASA, DM. Rx: cilostazol (arterial vasodilator), nifedipine (dihydropyridine Ca channel blocker) used in Raynaud's disease also. Tx: percutaneous balloon angioplasty or bypass. DDiagnosis: N root compress (sharp stabbing pain), spinal stenosis (weakness), DVT (pain at rest, tender, swollen, warm calf muscles), ↑RF: age, HTN, hyperlipidaemia, DM, tobacco, obesity, sedentary, prothrombotic condition).

BJGP April 2007 Targeted screening for peripheral arterial disease in GP: a pilot study in a high risk group. NC Campbell et al. ≥60yo screening Scotland. 364 attended. Prevalence lower than anticipated. ABPI ≤0.9, tobacco, diabetes, cholesterol.

BJGP Dec 2006: Symptomatic peripheral arterial disease: the value of a validated Q and a clinical decision rule. B Bendermacher et al. Netherlands. Observational study. 4790 patients with IC sxs. Prev PAD 48.3%.

Edinburgh claudication questionnaire (Do you get pain / discomfort in either leg on walking? Does it ever begin when standing or sitting? Does it occur if you walk at an ordinary pace on level? What happens if you stand still? In what part of your leg do you feel pain? inadequate diagnosis value. ABI required for diagnosis.

9. NICE 07: Reducing risk of DVT and PE in inpatients having surgery J Hill et al.

DVT 20% major ops and 40% major orthopaedic ops. PE 5%. RF for prophylaxis: > 60yo, BMI ≥ 30, continuous travel > 3h during 4 weeks pre/postop, immobility (cast, paralysis), personal or fhx of dvt, VV with phlebitis, active CA or CA tx, active heart or respiratory failure, severe infection, acute medical illness, recent MI or CVA, UC, Crohns, coc or HRT, pregnancy or puerperium, antiphospholipid syndrome, behcets disease, central venous catheter in situ, thrombophilias, myeloproliferative disease, nephrotic syndrome, paraproteinuria, paroxysmal nocturnal haemoglobinuria. Prevention: graduated compression or antiembolism stockings, intermittent pneumatic compression or foot impulse devices except if C/I (PAD or diabetic neuropathy). Offer mechanical prophylaxis and low MW hep to all orthopaedic op patients. Fondaparinux may be used as an alternative to LMWH. Offer thigh length stockings. Compression at ankle should be 18 mmHg, 14mmHg calf, 8 mm Hg upper thigh. Consider regional vs. GA. Leg exercises if immobile postop. Pros/ cons of stopping pre-existing anticoagulation tx preop. Stop coc 4/52 preop.

BMJ 31Mar 2007: Anticoagulation for 3 vs. 6 months in patients with DVT or PE or both: randomised trial. IA Campbell et al. 46 UK hospital 1 y FU. 3 vs. 6 months of heparin x5d then warfarin target INR 2-3.5. Little advantage to ↑ to 6/12 in patients with no persistent RFs and > risk of bleeding w/ longer duration warfarin tx.

BHF Feb 06 Factfile Air travel and venous thromboembolism 2-4 x↑risk with >10h duration. Compression stockings (15-30 mm Hg at ankle) ↓ risk (Lancet 2001 Scurr et al. Freq and prevention of symptomless DVT in long-haul flights: a RCT). Low MW hep for highest risk (s/e bleeding and hep-induced thrombocytopaenia). Aspirin has no effect (Angiology 02 Venous thrombosis from air travel: the LONFLIT3 study-prevention with aspirin vs. LMWH in high-risk subjects). Causal role of stasis, dehydration, cramped seats and hypobaric hypoxia. VTE 1/1000/y.

10. GP Mag 22 Sept 2006 Clinical review: AAA S Das.

Fhx 2x, M > 50 yo, tobacco, atherosclerosis, α1antitrypsin deficiency.
Elective repair 5-8% mortal vs. rupture 60%. MASS study screen 67,800 M 65-74 yo 53% risk reduction with screening, Lancet 2002, cost effective. < 5.5 cm small risk. EVAR (endovascular aneurysm repair) through groin incision and replace by graft reinforced to aortic wall with metallic stents. 1.7% 30d mortality vs. 4.7% open EVAR1. DREAM (Dutch) trial fast recovery. EVAR 2 showed no difference in overall survival between older high-risk patients (non-operable) vs. EVAR.

BMJ 19 April 2008 Should we screen for AAA? S Brearley, Whipps X, 5% M between 65 and 74 have AAA. Endovascular repair of AAA has mortality of 1.6% vs. 9.5% for open repair. DoH plans to screen all 65 yo M.

11. <u>BMJ 29 July 2006 Subarachnoid Haemorrhage</u> R Al-Shahi et al Suspect with sudden severe HA peaks w/n mins and lasts > 1h. Cannot exclude with un-enhanced CT so may need LP. Oral nimipodine 60 mg q 4h ↓ poor outcome. Endovascular coiling is superseding clipping for the occlusion of ruptured aneurysms.

12. <u>Acute intestinal ischaemia</u>

13. <u>Saphenous vein incompetence</u>

<u>BMJ 5 Aug 2006: Varicose veins and their mx</u> B Campbell. Convent op saphenofemoral ligation is the optimal tx for VV, cost effective. Laser and radiofrequency txs are alternatives to replace 1 part of traditional operations but most require additional phlebectomies or sclerotherapy. U/S (hand held Doppler and duplex) have replaced traditional tourniquet tests to assess VV. Sclerotherapy for small VVs below knee. Foam sclerotherapy for larger.

14. L shoulder pain in 30 yo M tennis player

DG: LH? Red flags. IPS: IADL, pro tennis s/e job £? Tried rx? ICE? OE: ROM. Hawkins-Kennedy test : 90° flex elbow in front of chest, force into IR pain = + impingement), empty can (Jobe's Test – arms extended out to side, thumbs point down and resist down pressure = impingement SS). AAFP article on shoulder exam. EdxCU: 1. supraspinatous tendinopathy/rotator cuff (SS, IS (resist ER), subscapularis = IR hands behind back and push out against your hand), teres minor = extend and abduct) painful arc >35 grad vs. < 35 trauma, hold deltoid in pain (referred pain from subacromial site) + impingement + drop arm (severe tear), pain lying on affected side 2. frozen = adhesive capsulitis (45-65 yo, ↓ER, then ↓abduction, then ↓IR) 3. AC joint sprain (<40 fall on tip of shoulder vs. older OA = painful horizontal adduction, AC compression (scarf) test) or 4. GHJ dislocation (deformity, not want to move, trauma RTA, rugby). CMx: Options: rest, lighter work, NSAID (stop if heartburn), steroid injection, 4-6/52 physio, scan vs. refer to orthopaedics. RGO. DTA. Modify work/ sports. CAB. PILS. FU.

BJGP Aug 2007: Shoulder adhesive capsulitis: system review of RCTs using multiple corticosteroid injection. N Shah et al. Beneficial up to 3 shots and up until 16/52 from date of 1st shot. Limited evidence 4-6. No evidence > 6.

15. BMJ 4 Nov 2006: Mobilisation with movement and exercise, corticosteroid injection, or wait and see for tennis elbow: RCT. L Bisset et al. Australia. 198 patients. Physio with elbow manipulation and exercise superior to wait and see in 1st 6 weeks, steroid injections after 6 weeks. S-t benefits of steroid injections reversed after 6 weeks.

16. Mechanical LBP vs. Sciatica 55 yo M

DG: Mechanical, nerve root pain, yellow flags vs. red flags. IPS: fear of 'slipped disc' vs. CA? Job? Lifts heavy boxes, Sports? Home? OE mechanical back pain vs. sciatica IPS: Empathy, ICE. Appreciate need to get back to work or play sports. EdxCU. Mechanical LBP vs. sciatica. Keep active. No x-ray or hosp referral if < 6/52 + no red flags. CMx: Options - OTC paracetamol, NSAID. LF. Warm up exercises. Refer for physio, orthopaedics (red flags), lighter work, ergonomics, The Back Pain Association (self-help), disability service team (return-to-work plan with employer). Nsaid/analgesia with PPI protection? or top Nsaid/analgesia with physio? PILS. HP: Weight advice. DTA. Warn red flags. RGO. Implications for sport, job.

Clinical Guidance mx of back pain, RCGP 1999 – paracetamol – NSAIDs – codydramol – baclofen/ diazepam. Back x-ray 120x radiation vs. cxr. 90% recover in 6/52.

BMJ 23 June 2007 Diagnosis and Tx of Sciatica BW Koes et al. 90% 2° to herniated disc w/ N root compression. Rare lumbar stenosis and CA. 5-10% patients with LBP have sciatica. Lifetime prevalence LBP 49-70%. Looked at systematic reviews. RFs: age (45-64), ↑ht, mental stress, tobacco, driving, lifting job. Ix: SLR or Lasegue's sign (sensitivity 91%, spec 26%), crossed SLR test (specificity 88%, sensitivity 29%). Diagnosis imaging if fail to respond to conservative tx for 6-8/52 or red flags. Disc herniation on CT/ MRI 20-36% in normal patients without sxs. Most respond w/n 2/52. NSAIDs v placebo 60% w/n 3/12. Stay active. Analgaesics or NSAID, acupuncture, epidural steroid injections, spinal manipulation, traction therapy, physical therapy, behavioural tx, multidisciplinary tx (unknown effectiveness for all). Surgical: removal of herniation; cauda equine emergency. 1.Paracatamol. 2. NSAID. 3. Tramadol. Paracetamol or NSAID in combo with codeine. 4. Morphine. No difference between conservative vs. disc surgery over 2 y FU just quicker relief. www.clinicalevidence.org

(BMJ), www.cochrane.iwh.on.ca (Cochrane back review group), www.backpaineurope.org (European commission research directorate general).

BJGP May 2006: Predicting persistent disabling LBP in general practice: a prospective cohort study. G Jones et al. Patients who report passive coping strategies (I wish my doctor would prescribe better pain med, depend on others for help with ADL, I can't do anything to lessen pain) have ↑ risk of persistent sxs and risk persists after controlling initial pain, severity and disability. Active coping (stay busy, distraction, physio).

17. BMJ 10 Mar 2007 Clinical review: Cervical spondylosis and neck pain. AI Binder. Clinical diagnosis: sxs: cervical pain aggravated by movement, referred pain (occiput, between shoulders, upper limbs), retro-orbital or temporal pain (from C1 or C2), cervical stiffness, numbness, tingling, or weakness in UL, dizzy, vertigo, poor balance, syncope, triggers migraine, 'pseudo-angina'. Signs: poorly localised tenderness, limited ROM (forward flex, backward extension, lateral flexion and rotation to both sides), minor neurological changes (inverted supinator jerks). Many > 30 have degenerative changes in IV discs with osteophyte formation and asymptomatic; ageing vs. disease? MRI for red flags (CA, infection or inflammatory-F, night sweats, unexplained weight loss, hx of inflammatory arthritis, CA, infection, Tb, HIV, drug dependency, IS, excruciating pain, intractable night pain, cervical LN, exquisite tenderness over a vertebral body; myelopathy – gait disturbance or clumsy hands, UMN signs in legs + LMN in arms, sudden onset in young pt suggests disc prolapsed; h/o severe osteoporosis, neck operation, drop attacks (vascular), intractable or ↑ pain. Rx: NSAIDs, amitryptylline, m relaxants, mobilisation physio. Radiculopathy (C5-C7) good prognosis, may respond to conservative tx. Neck operations for myelopathy/ pain.

18. Carpal tunnel syndrome 40 yo pregnant F. OE: Tinels and Phalens sign. Mx: nerve conduction studies, conservative night splint, steroid injection vs. refer for CTS release. PILS.

BMJ 18 Aug 2007: Clinical review: Carpal tunnel syndrome J Bland DDiagnosis: C6/7 radiculopathy, ulnar neuropathy, Raynaud's, vibration white finger, OA MCP thumb, de Quervain's tenosynovitis (Finkelstein test), MND, MS, syringomyelia (loss of Temp). Nerve conduction studies 5% FN. U/S 89% sensitivity, 69% specificity. Splint (37% satisfactory sx relief), steroids (70% initial response but relapse), surgical decompression 4000 patient survey 75% at 2 years.

19. Fever – sarcoidosis

20. Toe joint pain – gout
DG: psoriasis, hypothyroidism, ↑BP, renal disease, 1° hyperparathyroidism. Drugs: salicylates, thiazides. Surgery. IPS: ICE. Job? Married? EtOH red wine, port? Quail, grouse, tomatoes high in purine? Over-exercise? ↑ lactic acid, STI's (GC, urethritis), other joints? OE: red swollen painful big toe, BP, tophi ear. EdiagnosisCU. CMx: tests: uric acid, ?needle aspiration, diet, EtOH, stop salicylates or thiazides. rx indocin vs. naprosyn MR.

Pulse Mag 21 Dec 2006 Gout L Warburton GPSI. 70% 1st MTP, ankle, knee, wrist. ESR, CRP, sometime ↑wbc, platelets, FBG. 12% gout normal uric acid level. RF: male, hyperuricaemic, age, fhx, HTN, central obesity, EtOH, diuretics, trauma, renal insufficiency. Tx NSAIDs, colchicine (V/D). 40-80 mg methylprednisolone ankle or knee and 25 mg in MTP. Oral 20 mg prednisolone od for 4-5 days. Diet low purine. Long-term NSAID with allopurinol or can ppt acute attack. Take NSAID for 2 weeks when start allopurinol 50-300 mg of allopurinol to ↓urate to normal.

BMJ 3 June 2006: Diagnosis and mx of gout M Underwood. RF: diet-meat, fish, beer, spirits, obesity, weight gain. Drugs - diuretics, salicylates, ethambutol, cytotoxics, lead, pyrazinamide. Diseases - polycythaemia, chronic haemolytic anaemia, thalassaemia, G6PDD, Lesch-Nyhan syndrome, HTN, hypothyroidism, sickle cell anaemia, hyperparathyroidism, CRF.

GP Clinical 2 June 2006 The diagnosis and tx of gout. Prof R Sturrock. Can have normal uric acid. Do FBC to r/o polycythaemia. LFT, u/e to assess damage from raised uric acid. Inflammatory markers not helpful. After initial attack has subsided, 2 weeks, repeat uric acid, FBG and fasting cholesterol. Booklet from UK Gout Society.

Pulse 8 June 2006 M Fitzpatrick. Need to know Gout. Tx. Indomethacin, diclofenac or naproxen. RF (obesity, EtOH, purine-rich diet).Colchicine 0.5 mg tds better tolerated.

BMJ 9 Feb 2008 Mx of recurrent gout R Fox joint pain reaches max over 6-12h with swelling and erythema. Test blood for SUA and renal. Lose weight. www.ukgoutsociety.org. Reduce EtOH (beer rich in purine). If tophi or urate nephropathy present or has > 3 attacks a yr, consider allopurinol 100 mg 1 month after acute episode. ↑ by 100 mg q 4 weeks. Aim for SUA < 360 umol/l. Begin prophylaxis against acute attacks during 1st 3 months of allopurinol tx with colchicine 0.5 mg bd within 2 weeks before allopurinol. Warn risk of diarrhoea. If has no h/o urate stones, consider switch amlodipine to losartan as it has a mild uricosuric effect.

21. The Practitioner Nov 2006 Polymyalgia Rheumatica V Kyle et al. 50-59 yo F proximal girdle pain and EMS, fatigue, weight loss, T, loss of appetite. Transient peripheral synovitis of wrists, knees and SC joint. 25% have GCA (temporal) -HA, scalp tenderness, jaw claudication, visual. EMS > 1h, age > 50, bilateral upper arm tender. Ix: ESR>40/PV/ CRP, u+e, FBC, CK, Rheumatoid Factor, TFT, UA, LFT, serum electrophoresis. CMx: 15 mg prednisolone

od effective within few days 80%. Stop after 2 yrs in 50% and rest in 4 years. Small group need 2-3 mg od for years.
May add NSAIDs. DMARDs (methotrexate or azathioprine last resort). Refer GCA for biopsy + ↑ steroids.

22. <u>Pulse Mag 28 Sept 2006 Need to know SLE</u> T Myers. Joint pains (unlike RA swelling) are 'flu-like' – mild; photosensitive rash; fatigue and sl + ANA. More common in Afrocarribean and Asians. 9:1 F:M. APS (Hughes' syndrome) subset with + anticardiolipin antibody and thrombosis (causes epilepsy, CVA and recurrent miscarriage and 1 in 5 DVTs). Can teach patients to test urine for protein. ↑risk of MI and atheroma in 40s-50s. If F < 40, ANA, FBC, ESR, CRP, UA, not RhF. If ANA +, add antiDNA Ab and anticardiolipin (especially if thinking of pregnancy to exclude Hughes). Tx nephritis w/ cyclophosphamide or mycophenolate mofetil + steroids, possibly change to hydroxychloroquine, monitor q3/12 FBC, renal function, antiDNA, complement levels and UA. Overproduction of antibodies clogs blood vessels/ kidney filters. A negative ANA strongly against SLE, not r/o.

<u>GP Mag 27 Oct 2006 SLE</u> DP D'Cruz cons rheum Lupus Unit STH. Peak late teens and early 40s. Antiphospholipid syndrome (Hughes) livedo reticularis, recurrent miscarriages, renal artery stenosis, MT #, AVN. Requires long-term anticoagulation for thrombosis. SLE: joint 90%, intermittent, cutanenous rash 85%, splinter haemorrhages, 70% renal, pleural effusion, pericarditis, 30% systolic murmur, Libman-Sacks endocarditis complication, CNS lupus, abdo pain 20%, HSM 30%, normochromic normocytic anaemia of chronic dis. Platelets < 100. Persistent leucopenia < 4 classic. Anti-ds DNA, anti-Ro (SSA) antibody, anti-La (SSB) 20-30%. <u>Mx</u>: topical tacrolimus and pimecrolimus for discoid lupus lesions. NSAIDS joint pain. Hydroxychloroquine mild SLE arthralgia, oral ulcer, skin rash. Mepacrine safe antimalarial used with hydroxychloroquine. Ocular toxicity rare but annual check visual fields. Short course of low dose pulse cyclophosphamide then azathioprine for nephritis. mycophenolate mofetil for severe

nephritis, may supersede cyclophosphamide. Biological agents rituximab – long-lasting remissions after 2-4 infusions. Used already for lymphoma.

23. <u>BMJ 16 Sept 2006 Clinical review: Diagnosis and mx of ankylosing spondylitis</u> CM McVeigh et al. HLAB27. MRI SI joint early disease. Physio, regular NSAID, intraarticular steroid injections, sulphasalazine (inconclusive). New drugs inhibit TNFα (etanercept SC, infliximab IV and adalimumab SC, also used in severe RA) approved in Scot. Rapid clinical effects. Inflammatory back pain refer early to rheumatology. Diagnosis: LBP 3/12 with inflammatory characteristics (improved by exercise not rest). Limitation of L-spine motion in sagittal and frontal planes, ↓chest expansion, bilateral sacroiliitis ≥Grade 2, unilateral sacroiliitis ≥Grade 3. Morning stiffness > 30 mins. Alternating butt pain. Anterior uveitis (40%), aortic incompetency, pulmonary fibrosis, cardiac conduction abnormalities. ↑CRP, ESR. Bamboo spine (complete fusion). Osteoporosis and fractures common. Oral bisphosphonates (previous fracture). Op: THR, spinal surgery (fusion procedures for segmental instability and wedge lumbar osteotomy for fixed kyphosis.

24. <u>BMJ 16 Dec 2006 Clinical review: Osteoporosis and its mx</u> K Poole et al RF: age, previous fragility #, maternal hx of hip #, oral glucocorticoid tx, tobacco, EtOH ≥ 3U/d, RA, BMI ≤ 19. Depend on BMD: immobility. Malabsorption, endocrine, CKD, cirrhosis, COPD, aromatase inhibitors. Research: human monoclonal Ab to the R activator of NFkB ligand (RANKL) SC q 6mos, oral calciomimetic drugs that stimulate intermittent PTH, SERMs, inhibitors of sclerostin. WHO T score ≤-2.5. Ix: fbc, esr, LFT, renal, bone, Ig, paraproteins, urinary bence jones protein, tfts. Rx: bisphosphonates (alendronate 70 mg weekly, etidronate, ibandronate and rsiedronate); Strontium ranelate if C/I bisphoshonates. Teriparatide (PTH peptides) SC od. Raloxifine ↓vertebral #. HRT 2nd line. Ca and vitamin D with other txs. Fragility # start rx ASAP.

GP Mag 20 October 2006 Clinical review: Osteoporosis I Fogelman GKT med school. 1-3 mill UK. 25% F by 80 yo. 1in 3 > 50 vertebral # and 1 in 6 hip #. DM 2fold ↑risk#. RF: early menopause or hysteric (< 45 yo), FHx, tb, BMI < 19, long-term high dose steroids, comorbidities: thyrotoxicosis and hyperparathyroidism. Diagnosis: clinical features. #. Ix: measure BMD. DEXA scan WHO: normal T score > -1SD, osteopenia T score < -1 to > -2.5, osteoporosis T score < -2.5 SD. Tx NICE Jan 05: bisphosphonates inhibit bone resorption, ↓risk hip # from 25-50% and vertebral # 39-60%. UGI s/e. Strontium ranelate, raloxifine (SERM) but ↑risk DVT, hot flush, leg cramp. Teriparatide (recombinant PTH) anabolic effect on bone, expensive so for > 55. 2° prevention tx. 1° prevention > 75 yo.

25. Here for spine x-ray results for pelvic pain - Osteopenia– 70 yo slim F

DG: 'Nobody listens to me. I have pain in my bones.' Injuries? Home? Lives in warden controlled flat. Does not go out much. Isolated. No money for food. Lives on £46 a week to cover utilities and food PHx: R THR. OE: BMI 16, prominent tender pelvic bone. IPS: ICE? Fear of no money. Edx-CU: x-ray – osteopenic. CMx: Options: ix: dexa scan, bloods (fbc, bone profile), bisphosphonates , calcichew forte, diet vitamin D, oily fish, eggs, meat, vitamin D-fortified margarine or breakfast cereal. CAB. Community centre? Meals on wheels. Age Concern.

26. Mass in neck for 6 months – Multinodular Goitre 56 yo F

DG. OE: sweaty palms (hyperthyroid), coarse hair, pretibial myxoedema (hypothyroid), swallow with glass H2O – lump moves?, count to 10 to check for stridor, soft vs. firm thyroid, Trotter's sign (loss of laryngeal crepitus), thyroid bruit, percuss manubrium (retrosternal), Pemberton's sign, check hands, pulse, tremor. EdxCU: multinodular goitre and euthyroid. CMx: Indications for tx of a multinodular goiter: compression-induced sxs, hyperthyroidism, suspected CA, and cosmesis. Options: selective resection of pathologic tissue, bilateral subtotal thyroid lobectomy, and total thyroidectomy.

27. Interpretation of results – Conn's syndrome

DG: 46 yo M BP 176/112, muscle cramps, polyuria. Ix: Na 148, K 2.4, HCO3 34= met alkalosis. ECG: flattened T waves. EdxCU: 1° hyperaldosteronism, most likely Conn's (aldosterone secreting adenoma of adrenal cortex). CMx: Check AM aldosterone ↑, renin ↓. 2° = renin ↑. CT scan adrenals (CA, adenoma or hyperplasia). Conn's tx: spironolactone 4/52 before operation.

28. Exam of the diabetic foot

DG: 67 yo IDDM c/o numbness and burning in his foot, worse at night. IPS: ICE. OE: inspect for Charcot's joint, neuropathic trophic ulcers are over soles of feet over pressure points in diabetic neuritis vs. ischaemic painful ulcers, vascular + neurological exam, Doppler, absent ankle jerk reflex, stocking distribution of sensory loss, limited movement ankle and foot = diabetic neuropathy.

BMJ 14 July 2007 Effects of txs for sxs of painful diabetic neuropathy: systematic review M Wong et al. Anticonvulsants and antidepressants most commonly used. Oral TCA and traditional anticonvulsants better for s-t pain relief than newer. Long-term effects still lacking. Further studies needed on opioids, NMDA antagonist and ion channel blockers. Rx capsaicin if C/I to TCA → traditional anticonvulsant (sodium valproate, carbamazepine) →newer (pregabalin, gabapentin) → duloxetine → opioid.

Diabetes Care 2005 Under diagnosis of Peripheral Neuropathy in Type 2 DM WH Herman et al. GOAL A1C study.

29. Diabetes in pregnancy

30. Emergency mx of diabetic ketoacidosis

31. Diabetes BMJ 9 Dec 2006 Clinical review: Mx of hyperglycaemia in T2 diabetes RJ Heine et al. Affects 245 million worldwide. Metformin (↓ production of hepatic glucose) 1st line and ↓CVD risk. 2nd line add Sulfonylurea (stimulates insulin secretion): s/e weight gain, severe hypos especially in elderly on long-acting glibenclamide and chlorpropamide OR Thiazolidinediones (improves insulin sensitivity) less effective OR insulin (stimulates peripheral glucose uptake and inhibits hepatic glucose output) if HbA1C > 8.5. α-glucosidase (slows intestinal absorption of glucose) inhibitors (GI s/e). US: inhaled insulin: needs multiple daily dosing and expensive.

BMJ 10 Feb 2007: Pharmacological and lifestyle interventions to prevent or delay T2DM in people with impaired glucose tolerance: systematic review and meta-analysis. C Gillies et al. Lifestyle intervention (diet, exercise) at least as effective as_drug tx (antidiabetic drugs and antiobesity rx).17 trials in meta-analysis. Both ↓ progression.171 mill DM in world (2000) and estimate double in 2030. T2DM ↓life by 15 yrs = 5% NHS resources.

Diabetes Care 2004;27:155. XENDOS study supports use of orlistat in obese NIDDM.

32. BBN cancer or + HIV test

DG: Sxs? Ideas? Why test arranged? Job? Home? Husband, Children. IPS: tissue, touch. EdxCU. Warning shot. I'm sorry...We have all the time in the world. Prognosis. CMx: Options: sx relief. Urgent hospital referral (op/ chemo/ radiation), Macmillan support, PHCT. PILS. AC: DVLA? HIV: insurance, partner. Implications for job, family. FU with husband. Foster hope. Shall I arrange a taxi? Tea?

33. Terminal care – husband has met lung CA and wants to die at home

DG: Hospital setting, hospice vs. care at home. Advanced directives. What's in place? 24 hour PC nurse. DS1500 benefits. Disability living allowance. Painkillers, laxatives, anti-emetics. Night cover in place? Macmillan nurses. Green card scheme. Liverpool care pathways. Care of dying booklet. Relatives.

Pulse Mag 21 Dec 2006: Need to know Lung CA Prof Jessica Corner 3% of lung CA present with haemoptysis but 20% who have persistent haemoptysis have resectable lung CA. CT scan should be used to diagnosis prior to bronchoscopy for a suspicious CXR. PET scanning more sensitive in staging resectable CA. Stop smoking at 60, 50, 40, 30 gains 3, 6, 9 or 10 yrs of life. Stop < 35 same risk as non-smoker. Endobronchial radiotx or stenting in palliative care for breathing difficulties. Lung CA 20% Small cell (rx chemo). 80% NSCLC (squamous, adeno and large cell) – median survival 9 months and > 70% death w/n 1y.

Inform Benefits/ Entitlements: DS1500 (Practice benefits advisor – housing, incapacity benefit), DLA special regulations, Macmillan benefits advice line (website).

Apply for disability living allowance (£250/week) – will take 6/12 so do DS1500 also to get money faster to hear back in 2/52. Use this 'bribe' to get them to sign up to PC.

Ask carer – "Do you know what to do when they die?" Don't need to phone 999, if expected death. Call family members to visit. Call GP (no rush) at 9 am (may not visit until after surgery at 12). Contact undertakers during 9-5 (else charges more). Discuss hospital or not for each attack of COPD, renal failure, heart failure…

Support Groups for Carers or contact local CA network. National OOH guidelines has section on availability of drugs. PC drug formulary in pt's home and rx H2O injection as STAT dose or syringe driver in home:

Pain: WHO guidance: paracetamol 6x a day→codeine or tramadol→oramorph faster release→30 mg bd MST slow release (last or get side-effects). Morphine sulfate as national shortage of diamorphine.

Vomiting: cyclizine. Agitation: midazolam Psych: haloperidol IM. Chest secretions: glycopyrolate (not sedating) or buscopan (not cross BBB) antimuscarinic. NOT hyoscine because it crosses BBB and causes agitation and is more expensive.

Breathlessness in COPD: acute (lorazepam 0.5 mg to 1 mg SL) for chronic: use low dose morphine oramorph + benzo for anxiety component. Fan in face. Sit by open window. May transfer to Hospice for tx of breathlessness.

Antidepressant: citalopram, steroids (give energy boost, appetite stimulant) or methylphenidate (faster effect).

Unless you write a will, everything goes to the state. Info in local hospice – inheritance tax, pension. Family names need to be on tenancy or you will go homeless. Guardianship of young children. District nurses set up syringe driver. Prepared rx's. DN and PC nurses all do home visits.

Relatives have 24h access to PC nurse on phone. PCT can't find carers 24/7. If need that, then transfer to nursing home. Panicked carers/ relatives →unplanned hospital admission. Advanced CA: resuscitation does not work, nil. If you call ambulance, they will do CPR. End stage COPD: long-term oxygen; CRF – dialysis.

All doctors should be able to recognise the signs of patients entering the terminal phase (bed bound, unable to take sips of fluid, or take oral meds, semi-comatose).

Consider reversible causes of deterioration (hypercalcaemia, infection or opioid toxicity). In the terminal phase review the pt's meds and stop all non-essential meds.

All patients who intend to die at home should have a minimum of 3 days' supply of PC drugs. Anticipatory meds include diamorphine, midazolam, hyoscine, cyclizine and water for injections. This should be prescribed in advance as deterioration in a pt's condition can occur suddenly and is more likely OOH. All dying patients should have a catheter pack left in their home in case the OOHs service do not carry them. Carers are informed of what to do when the pt dies. The pt info sheet SCR2 is updated and copied to the OOHs providers. £2k simple cremation cost

Doctor confirms death before they can move the body + check all relatives visited. GP prompt visit if deceased in loo, in bed and partner shares, if family member alone. GP to tell when to pick up death cert and where it will be and to tell where is the registrars with directions. Leave advice written down as in shock. Burial (W Indies) vs. cremation (Hindu + Sikh). 7-10d time span to have funeral; social fund for cremation but need to be on housing or disability benefit.

<u>Social Services</u> – remove a dead body under environmental health act. Will bury person with no relatives. State fund to funeral directors. Priest will find someone to attend.

34. Demands subutex – 20 yo F heroin IVDA (based on my work at the drug misuse clinic, Hurley Clinic)

DG: Heard about subutex from friend. Drug hx, reason for starting (lost baby), quitting (boyfriend pressure), forensic hx (shoplifting), social hx (lives in squat), urine toxicology screen, HIV test, hepatitis testing + vaccination. Prior hx with drug rehab team – methadone did not work. IPS: Confidential. ICE subutex. Edx-CU: Discuss contract. CMx: Options: in-patient detox, CDPs (maintenance tx, needle exchanges), GPwSI. FU. Hep B vaccination.

BMJ 28 July 2007 Psychosocial interventions and opioid detox for drug misuse: summary of NICE S Pilling et al. Opioid detox (buprenorphine or methadone 1st line. Consider lofexidine for mild dependence. Do not use clonidine routinely ($\downarrow$BP), naltrexone , benzos or ultrarapid detox with GA. Psychosocial (empathy, opportunistic brief interventions, drug misuse services, voucher incentive, drug test, hep B, C, HIV, Tb test.

BJGP Nov 2006: Hep C infection among injecting drug users in GP: a cluster RCT of clinical guidelines' implementation. W Cullen et al. Ireland. 26 practices. Screen methadone maintenance tx patients for hep C and refer anti-HCV antibody + patients. Intervention group more likely to be screened vs. control but not SS referral results.

35. BMJ 7 July 2007: Clinical review: Managing smoking cessation P Aveyard et al. Meta-analysis all forms of NRT equally effective in long-term cessation. 5A's: ask about smoking, advise stop, assess motivation to stop and need for rx, assist with rx or refer to behave support programme, arrange FU. 1st line NRT, antidepressant (bupropion – 1:1000 seizure) licensed. Meta-analysis 31 trials OR 1.94 (1.72-2.19) or new drug varenicline (partial agonist act on

α4β2 nicotinic receptor. OR 3.22; 2.42- 4.27 to use with behave support. Nortriptyline 2nd line.

BJGP Oct 2006 A RCT of motivational interviewing for smoking cessation R Soria et al. Spain. Effectiveness at 6 and 12 months of motivational interview (3x20 min visits) 5.2 x higher than anti-smoking advice (1 visit).

BMJ 3 June 2006 ABC of COPD. Smoking Cessation. P Srivastava. 1° prevention of COPD-ads, bans, ↑price, smoke free policy at work/ public places, health warnings on packs, sustained health promotion campaigns. 2° prevention - behavioural support (set quit date, rid all cigarettes, partners quit together, FU), NRT, brief advice profs (2-3% quit long-term).

BMJ 9 Apr 05 Estimate of deaths attributable to passive smoking among UK adults: database analysis: K Jamrozik. Exposure at work 1/5 of all deaths from passive smoking aged 20-64 and up to 1/2 of deaths among employees of the hospitality industry.

BMJ 13 Mar 04 ABC of smoking cessation. Cessation interventions in routine health care. T Coleman. 5 'A's (ask at every opportunity, assess interest in stopping, advise against, assist to stop, arrange FU), H2Q (help2 quit) smoking cessation service in Shropshire.

NICE Mar 2006 SMOKING CESSATION

Brief interventions typically take between 5 and 10 mins and may include ≥ 1 of the following: simple opportunistic advice to stop, an assessment of the patient's commitment to quit, an offer of pharmacotherapy and/or behavioural support, provision of self-help material and referral to more intensive support i.e. the NHS Stop Smoking Services. Annual record. ↑165,000 more smokers to quit/year.

36. BJGP Dec 2006: Withdrawal from long-term benzodiapine use: randomised trial in family practice. C Vicens et al. Spain. Patients achieved withdrawal or reduced their doses by at least 50% after 6 and 12 mos. Standardised advice and tapering off schedule is effective and feasible in 1°care.

37. Alcoholism

DG: 50 yo noted to have ↑ MCV and GGT. Screen: CONTROL (control, neglect family, job, friends ask to reduce, drink in AM to overcome hangover) > 3U/d, DUI? Forensic hx? Depression? IPS: job? Army major. EtOH to relieve stress? CMx: AA, community alcohol teams, disulfiram (antabuse)? www.combatstress.org.uk.

BMJ Jan 05 Tx of EtoH related probs. B Ritson. Motivational interviewing - empathic non-confrontational. Benzos 1st line tx for withdrawal sxs and detox. If self-neglect, 200-300 mg thiamine/d x 3/12. Disulfiram deterrent but risk of hepatotoxicity so supervise and monitor LFTs in early mos. Acamprosate useful adjunct to psychological tx. Start as soon as abstinence achieved and use 1 year.

38. BMJ 24 Mar 2007: Clinical review: Managing acute organophosphorus pesticide poisoining DM Roberts et al. MI, USA. Sources: malathion (head lice), chlorpyrifos (cockroach bait), insecticides, room sprays, sarin, fumigation. Sx tx: saliva, lacrimation, N/V/D = atropine, glycopyrrolate; Bronchospasm = atropine, ipratropium, glycopyrrolate. ↓BP = fluids, atropine, vasopressors, inotropes. ↓HR = atropine, glycopyrrolate. Eye pain = mydriatics, cyclopegics. Muscle weakness = oximes. Respiratory failure = intubation, oximes. Seizures = benzodiapines. Acute cholinergic crisis – (muscarinic receptors) diarrhoea, urinary frequency, miosis, ↓HR, bronchorrhoea (focal creps or wheeze)/constriction, emesis, lacrimation, salivation, ↓BP. Nicotinic receptors - fasciculations, muscle weakness, respiratory failure, mydriasis, ↑HR, ↑BP. CNS-seizures, LOC Tx: 1-3 mg

atropine IV, double dose IV q3-5 mins until ↓HR and clear chest. 100s of mgs may be required in some. Oximes Pralidoxime chloride LD 30 mg/kg IV over 20 mins then continuous IV 8 mg/kg/h until 12h after stop atropine or once butyrylcholinesterase is noted to ↑ or use obidoxime. Benzodiapine IV start 5-10 mg diazepam (0.05-0.3 mg/kg/dose), lorazepam 2-4 mg (0.05-0.1 mg/kg/dose) or midazolam 5-10 mg (0.15-0.2mg/kg/dose). Decontamination: dermal spills (wash pt w/ soap and H20, discard clothes), gastric lavage if present w/n 1-2h, activated charcoal without cathartic 50g orally or NG if w/n 1-2h.

39. Insomnia/ Depression (NICE Oct 2009 DEPRESSION)

DG: 72 yo F lost her husband to CA, not sleeping well +EtOH, £ debt. Depression scale. Anhedonia. SI? IPS: bereavement, social isolation. Guilt over husband's suffering. EdxCU. Depression = insomnia. CMx: Options: watchful waiting, exercise, PHQ-9, ↓EtOH (depressant), counsellor, CBT, SSRI (mod-severe only – NICE). RGO. Open review. FU sooner if suicidal. Benefits advisor. Do not give benzos! 50:50 zopiclone. Ideally no rx.

BJGP Dec 2006 GPs' attitudes to benzodiapines and 'Z-drug' prescribing AN Siriwardena et al. 2005 84 postal Q W Lincolnshire PCT. Majority GPs attributed greater efficacy and lower s/e with Z drugs (zopiclone, zaleplon and zolpidem) vs. benzodiapines. Beliefs are not determined by current evidence or NICE.

BJGP Nov 05 Deciding who gets tx for depression and anxiety: a study of consecutive GP attenders. J Hyde et al. 439 consecutive GP attenders with diagnosis of depression, anxiety or both. GPs influenced by severity of sxs rather than their understandability in relation to recent life stresses or social context of distress. Men are more likely and those with anxiety d/o less likely to be offered active tx.

BJGP Nov 05 Qualititative study of an educational intervention for GPs in the assessment and mx of depression. L Gask et al. Interviewed 30 GPs in Liverpool and Manchester.3 barriers: GPs belief that he can impact on outcome, training and organisational context in which doctors had to implement what they learned. So start at medical school and postgraduate interventions should be tailored to the tx of depression and focussed at the level of organisation and not at the individual GP. Patients need more active role in mx of dep. CBT.

DEPRESSION (NICE OCT 2009)

Sxs should be present for at least 2 weeks and each sx should be present at sufficient severity for most of every day. DSM-IV Diagnosis Subthreshold depressive sxs: < 5 sxs of depression. Mild depression: few if any sxs in excess of 5 required to make the diagnosis and sxs result in only minor functional impairment. Mod depression: sxs or functional impairment are between mild and severe. Severe depression: most sxs and the sxs markedly interfere with functioning. Can occur with or without psychotic sxs.

Consider how the following factors may have affected the development, course and severity: any hx of depression and comorbid mental health or physical disorders, any past hx of mood elevation (to determine if the depression may be part of bipolar disorder); any past experience of, and response to, treatments; the quality of interpersonal relationships; living conditions and social isolation.

Step 1: Case identification and recognition Be alert to possible depression (particularly in patients with a past hx of depression or a chronic physical health problem with associated functional impairment) and consider asking 2 questions:
1.   During the last month, have you often been bothered by feeling down, depressed or hopeless?
2.   During the last month, have you often been bothered by

having little interest or pleasure in doing things?

<u>Step 2: For people with persistent subthreshold depressive sxs or mild to moderate depression</u>
<u>Advice on sleep hygiene</u>: establishing regular sleep and wake times, avoid excess eating, smoking or drinking alcohol before sleep, create a proper environment for sleep, taking regular physical exercise.

<u>Low-intensity psychosocial interventions</u> consider offering ≥ 1 of the following: individual guided self-help based on the principles of CBT, computerised CBT (CCBT), a structured group physical activity programme.

<u>Step 3: persistent subthreshold depressive symptoms or mild to moderate depression with inadequate response to initial interventions, and moderate and severe depression</u>: an antidepressant (normally a SSRI]) **or** a high-intensity psychological intervention option: CBT, interpersonal therapy (IPT), behavioural activation (but note that the evidence is less robust than for CBT or IPT), behavioural couples therapy for people who have a regular partner and where the relationship may contribute to the development or maintenance of depression, or where involving the partner is considered to be of potential therapeutic benefit.

<u>For moderate or severe, provide a combo of antidepressants & high-intensity psychological intervention (CBT or IPT).</u>
For patients who decline an antidepressant, CBT, IPT, behavioural activation and behavioural couples therapy, consider: counselling for patients with persistent subthreshold depressive sxs or mild to moderate depression, short-term psychodynamic psychotherapy for patients with mild to mod depression.

<u>Drug tx</u> SSRIs are associated with ↑ risk of bleeding, especially in older patients or those taking other rxs that have the potential to

damage the GI mucosa or interfere with clotting. Consider a gastroprotective drug in older patients taking NSAIDs or aspirin Fluoxetine, fluvoxamine and paroxetine are associated with a higher propensity for drug interactions than other SSRIs. Paroxetine is associated with a higher incidence of discontinuation sxs than other SSRIs.

Take into account toxicity in overdose when choosing an antidepressant for people at significant risk of suicide. Be aware that: compared with other equally effective antidepressants recommended for routine use in primary care, venlafaxine is associated with a greater risk of death from overdose. TCAs, except for lofepramine, are associated with the greatest risk in overdose. When prescribing rxs other than SSRIs, take the following into account: The increased likelihood of the patient stopping tx because of side effects (and the consequent need to increase the dose gradually) with venlafaxine, duloxetine and TCAs.

The potential for higher doses of venlafaxine to exacerbate cardiac arrhythmias and the need to monitor BP. The possible exacerbation of hypertension with venlafaxine and duloxetine. The potential for postural hypotension and arrhythmias with TCAs. The need for haematological monitoring with mianserin in elderly people. Dosulepin should not be prescribed. Non-reversible MAOIs, i.e. phenelzine, should normally be prescribed only by specialists.

For patients started on antidepressants who are not considered to be at increased risk of suicide, review after 2 weeks. See them regularly at intervals of 2 to 4 weeks in the first 3/12, and then at longer intervals if response is good.

A patient with depression started on antidepressants who is considered to present an ↑ suicide risk or is < 30 yrs (because of the potential ↑ prevalence of suicidal thoughts in the early stages of antidepressants for this group) should normally be seen after

<u>1 week</u> and frequently thereafter until the risk is no longer considered clinically important.

<u>If a patient with depression develops side effects</u> early in antidepressants, provide appropriate info and consider one of the following strategies: monitor sxs closely where side effects are mild and acceptable to the patient **or** stop the rx or change to a different antidepressant if the patient prefers **or** in discussion with the patient, consider short-term concomitant treatment with a benzodiazepine if anxiety, agitation and/or insomnia are problematic (except in patients with chronic symptoms of anxiety); this should usually be for no longer than 2 weeks in order to prevent the development of dependence.

<u>If response is absent or minimal after 3 to 4 weeks of treatment with a therapeutic dose of an antidepressant,</u> ↑ the level of support (i.e., by weekly face-to-face or telephone contact) and consider: ↑ the dose in line with the SPC if there are no significant side effects **or** switching to another antidepressant if there are side effects or if the patient prefers.

<u>When switching to another antidepressant, which can normally be achieved within 1 week</u> when switching from drugs with a short half-life, consider the potential for interactions in determining the choice of new drug and the nature and duration of the transition. Exercise particular caution when switching: from fluoxetine to other antidepressants, because fluoxetine has a long half-life (approx 1 week). When stopping an antidepressant, gradually reduce the dose over a 4-week period, although some may require longer periods, particularly with drugs with a shorter half-life (i.e. paroxetine and venlafaxine). This is not required with fluoxetine because of its long half-life.

<u>Do not use antidepressants routinely to treat persistent subthreshold depressive sxs or mild depression</u> because the risk–benefit ratio is poor, but consider them for people with: a past history of moderate

or severe depression or initial presentation of subthreshold depressive sxs that have been present for at least 2 years or subthreshold depressive symptoms or mild depression that persist(s) after other interventions.

Continuation and relapse prevention Support and encourage a pt who has benefited from taking an antidepressant to continue for at least 6 months after remission of an episode of depression. Discuss that this greatly reduces the risk of relapse and antidepressants are not associated with addiction.

Psychological interventions for relapse prevention Those at significant risk of relapse (including those who have relapsed despite antidepressants or who are unable or choose not to continue antidepressant therapy) or who have residual symptoms, should be offered one of the following psychological interventions: individual CBT for those who have relapsed despite antidepressant rx and for patients with a significant history of depression and residual sxs despite tx; mindfulness-based cognitive therapy for people who are currently well but have experienced ≥ 3 previous episodes of depression.

BJGP Aug 2007: Diagnosing depression in 1° care: using self-completed instruments: UK validation of PHQ-9 and CORE-OM. Simon Gilbody et al. 93 patients. PHQ-9 (sensitivity 91.7%, specificity 78.3%). CORE-OM (sensitivity 91.7%, specificity 76.7%) Brief self-rated Q is as good as clinical-admin instruments in detecting depression in 1° care.
BMJ 3 Feb 2007 Risk of suicide during tx with venlafaxine, citalopram, fluoxetine and dothiepin: retrospective cohort study A Rubino et al 219088 patients (18-89 yo) '95-'05. Main outcome measures: completed and attempted suicide. Venlafaxine had higher burden of RFs for suicide (previous attempts, severe depression). Venlafaxine vs. citalopram = (1.7; 0.76-3.8), fluoxetine = (1.63; 0.74-3.59) and dothiepin (1.31; 0.53-3.25) adjusted hazard ratios.

40. <u>Panic attacks (Panic disorders NICE 2004 ANXIETY)</u>

<u>DG</u>: 45 yo F. Job? Home? Caffeine. Drugs. Hyperthyroid? Pheo? EtOH? OE: BP/P, cvs, thyroid. <u>IPS</u>: stress, ICE <u>EdxCU</u>. <u>CMx</u>: Options: relaxation training, rebreathing paper bag, CBT, propranolol, wait and see, self help diary, Ix: TFT. FU. RGO. DTA. Less stressful job - change job? Holiday?

<u>NICE ANXIETY DISORDERS – Dec 2004: Tx in 1° Care for Panic disorders</u> Benzos are assoc with a less good outcome in the long term and should NOT be prescribed for tx. Sedating antihistamines or antipsychotics should NOT be prescribed. Interventions in order of longest duration od effect: CBT (7-14h total), SSRI licensed for panic disorder for at least 6 months after show improvement (or imipramine or clomipramine if no improvement after 12 weeks or SSRI not suitable), self help (biblotherapy, review between every 4 and 8 weeks).

<u>BMJ 17 Mar 2007: Clinical review: Generalised anxiety d/o</u> C Gale et al. Excessive and inappropriate anxiety and worry about many events or thoughts. Most also have other mood and anxiety d/o. CBT I more efficacious than non-directive psychotherapy or no tx. Anxiety mx tx also better than no tx. Antidepressants, benzodiazepines, buspirone and kava (hepatic compromise) are efficacious but have clinically significant adverse effects.

<u>BJGP Nov 05: Tx of anxiety disorders in primary care practice: a RCT.</u> C Boeijen et al. Dutch. Primary care patients with prevalent anxiety d/o for whom the GP does not find referral necessary, can be adequately tx by GP. Psych OPC referral does not give superior results. Guided self-help is easier for GPs to carry out.

41. <u>BMJ 26 Aug 2006. Clinical review: OCD</u> I Heyman et al Most often in adolescence. Sxs: Obsessions (fear of causing harm to self or others, contamination, need for symmetry or exactness, sex or religious obsessions, fear of behaving unacceptably, fear of making

a mistake). Compulsions (behaviours- clean, hand wash, checking, order and arrange, hoarding, asking for reassurance; mental acts – counting, repeat words silently, ruminations, neutralising thoughts). ICD-10 most days for at least 2 weeks. Repetitive, unpleasant, unreasonable, originating from patient's mind, at least 1 unsuccessfully resisted, carrying out thought or compulsive act is not intrinsically pleasurable. Quick screen: do you wash or clean a lot? Do you check things a lot? Is there any thought that keeps bothering you that you would like to get rid of but can't? Do your daily activities take a long time to finish? Are you concerned about orderliness or symmetry? Do these problems trouble you? Mild: self help/ info for family. Moderate-Severe: CBT Severe: CBT + citalopram or clomipramine.

42. PTSD DG: 25 yo M can't sleep. Been abroad. Sx? IPS: job? Iraq. CMx: refer charity www.combatstress.org.uk.

NICE March 2005 Definition: following a stressful event or situation of an exceptionally threatening or catastrophic nature, likely to cause pervasive distress in almost anyone. 25-30% experiencing trauma develop PTSD.

Sxs: re-experiencing sxs, flashbacks, nightmares, repetitive and distressing intrusive images or other sensory impressions. Reminders arouse intense distress and/ or physiological reactions. Avoidance of reminders of the trauma. Avoid thinking about the event or ruminate excessively as to why the event happened. Sxs of hyperarousal (hypervigilance for threat, exaggerated startle responses, irritability, difficulty concentrating and sleeping problems); emotional numbing (amnesia, lack of ability to experience feelings, feeling detached from others, giving up previously significant activities); anger.

Events: RTA, assault, rape, childhood sex abuse, traumatic childbirth, domestic violence. Comorbidities: depression, drug and alcohol, personality d/o, suicide/ self-harm. Initial response to trauma: Systematic provision to that individual alone of brief, single-session interventions (debriefing) that focus on the traumatic incident should NOT be routine practice when delivering services. Where sxs are mild and present for < 4 weeks after the trauma, consider watchful waiting. FU w/n 1 month.

Trauma-focused psychological therapies where sxs are present w/n 3 months of a trauma. Trauma-focused CBT (8-12 sessions) offer to those with severe PTSD in the 1st month after the traumatic event on an individual outpatient basis. All with PTSD should be offered a course of trauma-focused psychological treatment (trauma-focused CBT or eye movement desensitisation and reprocessing (EMDR)) on an individual outpatient basis. Drug tx may be considered in the acute phase of PTSD for mx of sleep disturbance - short-term hypnotic. For long-term, consider antidepressant. Non-trauma-focused interventions i.e. relaxation or non-directive therapy should NOT routinely be offered w/n 3 months of the event.

Drug treatments for adults (not to be prescribed for children and young people) Should NOT be used as a routine 1st-line tx in preference to a trauma-focused psychological therapy. Paroxetine or mirtazapine for general use and amitryptiline or phenelzine for initiation only by MH specialists should be considered in adults who express a preference not to engage in psychological treatment. ↑ the dose if no response to drug tx. If further drug therapy is considered, it should be with a different class of antidepressant or involve adjunct use of olanzapine. If responding, cont for 12 months before gradual withdrawal.

Adult PTSD with ↑ suicide risk and all between 18 and 29 (because of the potential ↑ risk of suicidal thoughts associated with use of antidepressants in this age) should normally be seen after 1 week and frequently thereafter until the risk is no longer considered significant. In the initial stages of SSRI, GPs should actively seek out signs of akathisia, suicidal ideation and ↑ anxiety/

agitation. Marked +/ or prolonged akathisia - review use of drug. If not at ↑ risk of suicide, review at 2 weeks and at intervals of 2-4 weeks in the 1st 3 mos. Reduce dose over 4-weeks.

Screening for PTSD: routine use of brief screening instrument for PTSD at 1 mo after a major disaster should be considered for disaster plan. Programme refugees and asylum seekers.

BMJ 14 April 2007 Clinical Review: PTSD J Bisson. Re-experiencing phenomena (x1), avoidance and numbness (x 3), ↑d arousal (x 2). Tx (exposure therapy, trauma focused cognitive therapy (w/n 3/12 tx of choice) or eye movement desensitisation and reprocessing for chronic). Paroxetine, mirtazapine 2nd line.

BJGP May 2006: PTSD in 1° care with special reference to personality d/o comorbidity M Gomez-Beneyto et al.Women who experienced high frequency body-contact traumatic events at an early age often suffer from personality d/o and present a particularly severe form of PTSD deserving referral to 2⁰ care.

43. Domestic violence in a 40 yo Oriental F occurs in 1 in 4 F!

DG: 'pain in my chest' (heartache) Injury? How? Gambling husband, 4 children. Poverty. Criminal record? Examine. Children at risk? IPS: You look sad. ICE. EdxCU: DV. CMx: Options: Support? Call 24h national DV helpline, women's refuge. Risk assessment. Plan escape route. Police to obtain injunction? Refer counsellor. Duty SW +HV if signs of child abuse.

PILS: Karma Nirvana 01332 604098, IKWRO (Iranian and Kurdish Women's Rights Organisation: 0207 490 0303 www.ikwro.org.uk, forced marriage unit 0207 008 0151 or www.fco.gov.uk/forced marriage, Victim supportline 0845 303 0900 www.victimsupport.org, national domestic violence helpline 0808 2000 247 www.refuge.org.uk. FU 3/7 to discuss plan. Help for husband? CAB.

BMJ 23 July 2007: More vigilance needed to tackle domestic violence. 3 /10 F and 2 /10 M suffer DV in their life (4 types – phys, sexual, psychological and financial) Health problems: fractures, burns, depression, PTSD, chronic pain syndrome, arthritis, hearing and sight problems, seizure, ulcer, heart disease, HTN. 80% F and 30% starts in pregnancy. BMA's Board of Science recommends more training to how to deal with DV.

BJGP April 2006: Recognising Domestic Violence in clinical practice using the diagnoses of PTSD, depression and low self-esteem. F Duxbury. Fear of health profs asking about trauma. Use diagnostic frameworks for depression and PTSD to gain trust and detect DV. ↑ IBS, chronic pelvic pain and gynae problems.

Increased awareness of intimate partner abuse after training: a RCT. S Wong et al. 37-41% of women have reported abuse from partner. Training most significant determinant to improve awareness and identification of partner abuse. Active questioning ↑ especially where there were non-obvious signs.

The acceptability of routine inquiry about DV towards women. A Boyle at al. Most acceptable to women who have not been abused in last yr and are attending ANC.

BMJ 7 April 2007: Intimate partner violence LE Ferris WHO's study on DV in 24,000F, 10 countries. Need for immediate action. Train health professionals to recognise, provide services, coordinate sectors, add anti-violence to antenatal, parenting classes, men's health services.

44. BMJ 2 June 2007 Clinical review: Rape and sexual assault J Welch et al. STIs (GC, syphilis, chlamydia, TV, HIV, hepatitis B) Mouth sample sperm lasts 31h. Pregnancy rate from rape 5%. L-2 rx. PEP HIV: HIV+ rapist, rapist RFs (high prevalence area, IVDA), assault within 72h, anal rape, trauma and bleeding, multiple rapists).

www.rcne.com (rape crisis network Europe – info on counselling, legal and support services in 30 countries), www.careandevidence.org (DVD training on collecting evidence) www.uktrauma.org.uk, www.rapecrisis.co.uk, www.met.police.uk/sapphire.advice/htm, www.victimsupport.org.

45. Weight loss – eating disorder/ anorexia (NICE Eating Disorders Jan 2004)

DG: 18 yo F weight loss, amenorrhoea and depression, exercise, sleep, binge eating vs. induce vomiting, FH psychiatric illnesses, menses, suicide, drugs, alcohol, tobacco. Job: librarian – no stress. OE: lanugo hair, burns in oral cavity, ketone breath. CMx: blood tests to rule out organic cause. Mx: psychotherapy or SSRIs.

BMJ 28 April 2007 Clinical review: Anorexia nervosa J Morris et al. average age onset age 15. Highest mortality of any psych d/o. ICD-10 criteria all 5: Body weight 15% < expected or BMI ≤ 17.5; self-induced by avoid fatty foods, purge, excess exercise, appetite suppressants or diuretics; distorted body image; endocrine d/ o (amenorrhoea (except if on coc), GH and cortisol may be ↑, thyroid, abnormal insulin secretion); arrested puberty. Compulsory tx by MH legislation in acute emergency where patient is unable to accept treatment. Tx: individual (cognitive analytic therapy, CBT, interpersonal psychotherapy, motivational enhancement tx, dynamically informed tx), group tx, family work. Takes 5-6 yrs from diagnosis to recovery. www.edauk.com.

Initial assessment (physical, psychological and social needs, assessment of risk to self) and the initial coordination of care, including determine need for emergency medical or psychiatric assessment. Supply patients and carers with info on self-help groups and support groups. Patients who vomit should have regular dental reviews and give advice: avoid brushing after vomiting; rinse with a non-acid mouthwash after vomit and reduce acid oral environment (limit acidic foods). Refrain from physical activities that significantly ↑ falls. At risk of osteoporosis.

Care of children (aged 8 or older)/ adolescents Family interventions should be offered. Monitor growth and development - paediatric advice. Alert to indicators of abuse (emotional, physical, sexual). Respect right to confidentiality.

Identification and screening in 1° care Young F with low BMI. Patients consulting with weight concerns who are not overweight. F with menstrual disturbances or amenorrhoea. Patients with GI sxs. Patients with physical signs of starvation or repeated vomiting. Children with poor growth. Screening: 'Do you think you have an eating problem?' and 'Do you worry excessively about your weight?' Young patients with Type I diabetes and poor tx adherence.

Assessment and mx of Anorexia nervosa in 1° Care Weight and BMI should not be sole indicators of physical risk as may be unreliable. Overall clinical assessment repeated over time (rate of weight loss, child's growth rate, objective physical signs, lab tests). Annual physical and mental review by GP unless under care of a 2° service. Be aware of ↑ risk of self-harm and suicide at times of transition between services or service settings.

Psychological interventions: Cognitive analytic therapy (CAT), CBT, interpersonal psychotherapy (IPT), focal psychodynamic therapy and family interventions explicitly on eating disorders. Outpatient psychological tx for 6 months with physical monitoring. If deteriorating, move from individual to combined individual and family work; or day-care or in-care (significant risk of suicide or self-harm, high or moderate physical risk). Following in-patient weight restoration, should have 12 months of outpatient psychological treatment.

Drug interventions Limited evidence. Caution as compromised CV function. Risk of drugs that prolong the QTc interval, ex antipsychotics, TCA, macrolide antibiotics and some antihistamines. If on rx, then ECG monitoring. All patients with diagnosis of anorexia should have alert on prescribing record concerning risk of S/Es.

Physical mx Average weekly gain of 0.5-1kg in inpatients and 0.5 kg in outpatient. Requires an additional 3500-7000 cals a week. MVI/multi-mineral supplement in oral form. Inpatients need skilled implementation of refeeding with careful physical monitoring (1st few days of refeed) in combo with psychosocial interventions. TPN should NOT be used unless there is significant GI dysfunction. Oestrogen administration should NOT be used to treat bone density problems in children/ adolescents as this may lead to premature fusion of the epiphyses. Feeding against the will of the patient requires expertise in the care and mx of those with severe eating d/o's and the physical complications associated with it. This should be done in the context of the MH Act 1983 or Children Act 1989 (the right of those with parental responsibility to override the young person's refusal). The legal basis for such an action must be clear.

Bulimia nervosa Psychological interventions. Encouraged to follow an evidence-based self-help programme. Offer CBT-BN (specifically-adapted form of CBT) to adults. 16-20 sessions over 45 months. Consider IPT as alternative to CBT, but inform patients it takes 8-12 months to achieve results comparable with CBT.

Drug interventions: As an alternative or additional 1st step to using self-help, adults may be offered a trial of an antidepressant. Advise that antidepressants can ↓ the frequency of binge eating and purging, but the long-term effects are unknown. Any beneficial effects will be rapidly apparent. SSRIs (fluoxetine 60 mg od) are the DOC for the tx of BN. No other drugs are recommended.

Mx of physical aspects of BN: Assess fluid and lyte balance (vomiting/laxatives). Oral supplementation. BN with poor impulse control (notably substance misuse) less likely to respond to a standard programme of treatment. For all eating disorders – Family members (including siblings) should be included in the treatment of children and adolescents. Share info, advise on behavioural mx and facilitate communication.

46. BMJ 14 July 2007 Clinical review: Schizophrenia M Picchioni et al. KCL.

Positive sxs: lack of insight, hallucinations, delusions, thought disorder. Negative sxs: social withdrawal, neglect, emotional blunting, and paucity of speech. NICE treatment: amisulpiride, risperidone, quetiapine or olanzapine low dose, titrate and reassess over 6-8 weeks. Continue, change or consider both typical and atypical or if poor compliance, consider depot or compliance therapy. Last resort clozapine.

Refer if poor treatment compliance, treatment response, ongoing substance misuse or ↑ in risk profile.

S/E 1st generation EP effects: dystonia, pseudoparkinsonism, akathisia, tardive dyskinesia; sedation, ↑prolactin, postural hypotonia, anticholinergic, neuroleptic malignant syndrome, weight gain, cardiotoxicity (prolonged QTc). 2nd gen olanzapine (weight gain, sedating, glucose intolerance and DM, hypotonia), risperidone (↑PRL, ↓BP, EP high dose, sexual dysfunction), amisulpiride (↑PRL, insomnia, EP), quetiapine (↓BP, dyspepsia, drowsy), clozapine (sedating, ↑saliva, ↓BP and ↑BP, HR, pyrexia, weight gain, ↓seizure threshold, glucose intolerance, DM. weight gain, pyrexia.

BMJ 31 Mar 2007 Clinical review: Managing the acute psychotic episode P Byrne.

Ix: urine toxicology screen, UPT, fbc, urea/lytes, random glucose, LFTs, Ca, TFTs cortisol if endocrine sxs. HIV high risk, ECG, EEG (temporal lobe epilepsy), CT/MRI.

Rx: antipsychotic at lowest dose + sedate with benzo.Typical antipsychotics: chlorpromazine 200 mg od (weight gain, EP s/e, sedation), haloperidol 2 mg od (poorly tolerated), pimozide 4mg od, sulpiride 400 mg, trifluoperazine 10 mg (EP s/e), zuclopenthixol acetate (like haldol). Atypicals: amisulpride 400 mg (fewer s/e), aripiprazole 15 mg, clozapine 250-550 mg (31 studies show is best effect), olanzapine 5 mg (fewer EP s/e, weight gain+), quetiapine 150 mg (fewer EP s/e, +sedate), risperidone 2 mg (not as good as clozapine), zotepine 75 mg (fewer EP side-effects).

Positive Psychotic sxs (paranoid delusion, grandiose, thought interference, passivity, thought echo, 3rd person auditory hallucinations; less clear 2nd person auditory hallucinations (voices speaking to pt – you're useless), thought disorder (thought block, overinclusive thinking (unnecessary detail), difficulties in abstract thinking); Psychotic sxs (apathy, blunted affect, emotional withdrawal, flat affect, odd or incongruous affect, lack of attention to personal hygiene, poor rapport, lack of spontaneity + flow of conversation). 1+ or 2 negative sxs = schizophrenia.

DDx: drug, schizophrenia, bipolar, schizoaffective, delusional, severe depression, PTSD, OCD, paranoid personality disorder, Asperger's syndrome, ADHD. Fhx of mental, suicide, EtOH, drugs, endocrine, EP side-effects antipsychotics (early akathesia, dystonic reactions, rare fatal neuroleptic malignant syndrome, late tardive dyskinesia).

47. BJGP Sept 05 Diagnosis and mx of patients with bipolar disorder in primary care. M Berk et al. Melbourne.

WHO ranks bipolar as world's 6th leading cause of disability-adjusted life yeas among 15-44 yo. 25-50% attempt suicide. Lithium ↓ risk of suicide and is used for both acute and prophylactic. Valproate for mania. Carbamazepine for long-term maintenance. Lamotrigine for acute and long-term bipolar. Atypicals (olanzapine, risperidone, quetiapine) - acute mania and early evidence for bipolar and olanzapine in maintenance.

48. <u>Chronic fatigue syndrome (ME)</u> DG: TATT. IPS: ICE? Diagnosis criteria – debilitating fatigue causing ↓ in activity to < 50% x 6/12 + sxs not explained by other medical or chronic psych illness. Sxs (4 required) – sore throat, painful axillary or cervical LNs, m discomfort/ pain, prolonged fatigue after exercise, HA, forgetful, un-refreshing sleep. IPS: IADL, job, family? EdxCU. Option: graded exercise/ CBT. PILS.

NICE CHRONIC FATIGUE SYNDROME Aug 2007

Consider CFS/ME if a patient has: Fatigue with ALL of the following features: new or had a specific onset (not lifelong), persistent and/or recurrent, unexplained by other conditions has resulted in a substantial ↓ in activity level, characterised by post-exertional malaise and/or fatigue (typically delayed by at least 24 hours, with slow recovery over several days) and ≥ 1 of the following symptoms: difficulty with sleeping (insomnia, hypersomnia, unrefreshing sleep, a disturbed sleep–wake cycle); muscle and/or joint pain that is multi-site and without evidence of inflammation; headaches; painful lymph nodes without pathological enlargement; sore throat; cognitive dysfunction (difficulty thinking, inability to concentrate, impairment of short-term memory, and difficulties with word-finding, planning/organising thoughts and info processing); physical or mental exertion makes symptoms worse; general malaise or 'flu-like' symptoms; dizziness and/or nausea; palpitations in the absence of identified cardiac pathology. Be aware that the sxs of CFS/ME fluctuate in severity and may change in nature over time.

Signs and sxs that can be caused by other serious conditions ('red flags') should not be attributed to CFS/ME without consideration of alternative diagnosis or comorbidities. The following features should be investigated: localising/focal neurological signs; signs and sxs of inflammatory arthritis or connective tissue disease; signs and sxs of cardiorespiratory disease; significant weight loss; sleep apnoea; clinically significant lymphadenopathy.

Hx (exacerbating and alleviating factors, sleep disturbance and intercurrent stressors) and physical and assessment of psychological wellbeing should be carried out. A child who has sxs suggestive of CFS/ME should be referred to paeds for assessment to r/o other diagnosis w/n 6 weeks of presentation.

Ixs: UA for protein, blood and glucose; FBC; urea and lytes; LFTs: TFTs; ESR or plasma viscosity; CRP; random blood glucose; serum creatinine; screening blood tests for gluten sensitivity, serum calcium; creatinine kinase assessment of serum ferritin levels (children and young people only). Serum ferritin in adults should not be carried out unless a FBC and other haematological indices suggest Fe deficiency. Vitamin B_{12} deficiency and folate levels should not be carried out unless FBC and MCV show a macrocytosis.

Serological testing should not be carried out unless the hx is indicative of an infection. Depending on the hx, tests for the following infections may be appropriate: chronic bacterial infections, i.e. borreliosis; chronic viral infections, i.e. HIV or hepatitis B or C (BBC news: diagnosis of hepatitis C was missed in a pt with CFS!); acute viral infections, i.e. infectious mono (use heterophile antibody tests); latent infections, i.e. toxoplasmosis, EBV or cytomegalovirus.

Advice on sx mx should not be delayed until a diagnosis is established. This advice should be tailored to the specific symptoms the person has, and be aimed at minimising their impact on daily life and activities. If sxs do not resolve as expected in a patient initially suspected of having a self-limiting condition, listen carefully to the patients and their family and/or carers' concerns and be prepared to reassess their initial opinion.
Consider discussion with a specialist if there is uncertainty.

A diagnosis is made after other possible diagnoses have been excluded and the sxs have persisted for: 4 months in an adult or 3 months in a child or young person; the diagnosis should be made or confirmed by a paediatrician.

Realistic info about CFS/ME and encourage cautious optimism. Most will improve over time and some will recover and be able to resume work and normal activities. However, others will continue to experience sxs or relapse and some with severe CFS/ME may remain housebound. The prognosis in children and young people is more optimistic.

The diagnosis of CFS/ME should be reconsidered if none of the following key features are present: post-exertional fatigue or malaise; cognitive difficulties; sleep disturbance; chronic pain.

<u>Sx mx</u> - There is no known drug tx or cure for CFS/ME. However, sxs should be managed as in usual clinical practice. If patients with CFS/ME have concerns, consider starting drug tx for CFS/ME symptoms at a lower dose than in usual clinical practice. The dose may be increased gradually, in agreement with the pt.

Although exclusion diets are not generally recommended for managing CFS/ME, many find them helpful in managing sxs, including bowel sxs. If undertaking an exclusion diet or dietary manipulation, seek advice from a dietitian because of the risk of malnutrition.

Provide tailored sleep management advice that includes: Explaining the role and effect of disordered sleep or sleep dysfunction in CFS/ME. Identifying the common changes in sleep patterns seen in CFS/ME that may exacerbate fatigue sxs (insomnia, hypersomnia, sleep reversal, altered sleep–wake cycle and non-refreshing sleep). Providing general advice on good sleep hygiene. Introducing changes to sleep patterns gradually.

If sleep mx strategies do not improve the person's sleep and rest, the possibility of an underlying sleep disorder or dysfunction should be considered, and interventions provided if needed.

Sleep mx strategies should not include encouraging daytime sleeping and naps. Rest periods are a component of all mx strategies for CFS/ME. Advise people with CFS/ME on the role of rest, how to introduce rest periods into their daily routine, and the frequency and length appropriate for each patient. Limit the length of rest periods to 30 minutes at a time. Introduce 'low level' physical and cognitive activities (depending on the severity of sxs).

Relaxation techniques should be offered for the mx of pain, sleep problems and comorbid stress or anxiety. There are a number of different relaxation techniques (guided visualisation or breathing techniques) that can be incorporated into rest periods.

Pacing helpful in self-managing CFS/ME. Advise that, at present, there is insufficient research evidence on the benefits or harm of pacing. Well-balanced diet in line with 'The balance of good health.' They should work to develop strategies to minimise complications that may be caused by nausea, swallowing problems, sore throat or difficulties with buying, preparing and eating food. Eating regularly, and including slow-release starchy foods in meals and snacks. For moderate or severe CFS/ME, providing or recommending equipment and adaptations (wheelchair, blue badge or stair lift) should be considered as part of an overall mx plan, taking into account the risks and benefits for the individual. This may help them to maintain their independence and improve their quality of life.

Having to stop their work or education is generally detrimental to patient's health and well-being. Therefore, the ability to continue in education or work should be addressed early and reviewed regularly. Proactively advise about fitness for work and education, and recommend flexible adjustments or adaptations to work or studies to return to them when they are ready and fit enough. This may include, with the informed consent of the pt, liaising with employers, education providers and support services, such as: OH services, disability services through Jobcentre Plus, schools, home education services and local education authorities, disability advisers in universities and colleges.

For patients who are able to continue in or return to education or employment, ensure, with the patient's informed consent, that employers, occupational health or education institutions have info on the condition and the agreed mx plan.

<u>Referral to specialist CFS/ME</u> care should be offered: w/n 6 months of presentation to patients with mild CFS/ME
- within 3–4 months of presentation to patients with moderate CFS/ME sxs
- immediately to patients with severe CFS/ME sxs.

<u>Specialist CFS/ME care</u> After a patient is referred, an initial assessment should be done to confirm the diagnosis. If general mx strategies are helpful, these should be continued after referral to specialist CFS/ME care. CBT, <u>graded exercise therapy</u> and activity mx programmes. Prescribing of low-dose TCAs, specifically amitriptyline, should be considered for people with CFS/ME who have poor sleep or pain. Should not be offered to people who are already taking SSRIs because of the potential for serious adverse interactions. Melatonin may be considered for children and young people with CFS/ME who have sleep difficulties, but normally under specialist supervision because it is not licensed in the UK. <u>Advise that setbacks/relapses are to be expected</u> as part of CFS/ME.

49. <u>Facial nerve palsy</u>

50. <u>20 yo F LOC – Diagnosis and Discharge instructions for epilepsy (NICE 2004)</u>

1
2
3

<u>DG</u>: witnessed, aura, hallucinations, incontinence of urine and sore tongue. <u>IPS</u>: ICE. Home? Job? Drives? <u>EdxCU.</u> Avoid strobe lights, EtOH, excessive tiredness. Wear alert bracelet. Swim accompanied and bathe – unlocked door. <u>CMx:</u> Options: refer. DVLA (Sept 09) 6 months fit free. ↑ dose of coc/ pop. FU. Support groups. PILS.

<u>NICE EPILEPSY October 2004</u>

GPs to refer patients < 2 weeks after 1st suspected seizure. Early referrals as 15-30% of patients are misdiagnosed. Diagnosis established by specialist. An EEG is only performed to support a diagnosis of epilepsy. MRI is the imaging ix of choice and indicated for new onset epilepsy < age 2 or in adulthood, medication resistant seizure, and for patients with a focal onset suggestive on hx, exam or EEG. Ixs: serum lytes, glucose, calcium, 12-lead ECG and no longer include prolactin. Yearly review by GP. Info on SUDEP to patients. 1st line: sodium valproate, carbamazepine, oxycarbazepine, lamotrigine, topiramate, ethosuximide. Tx cessation if seizure-free for 2 years.

<u>Drug tx of epilepsy in women on the pill</u> Avoid hep enzyme inducers (carbamazepine, oxycarbazepine, phenytoin, phenobarb, topiramate) if on coc, pop. Offer depo-provera q 10/52 instead of 12 or offer non-hormonal IUD. If taking enzyme-inducing AED and elect to take the coc, then a minimum of 50 mcg of oestrogen is recommended. For BTB, ↑ oestrogen to 75-100 mcg or tricycle 3 packs (no breaks). POP and the progesterone implant not recommended. If EC is prescribed, then 1.5 mg of levonorgestrel is required initially, then 750 mcg.

<u>Drug tx of epilepsy in pregnant women</u> Avoid carbamazepine, phenytoin, sodium valproate - shown to be linked with neural tube defects. Prescribe 5 mg folic acid daily preconceptually and during the 1st trimester. Recommend use lamotrigine or gabapentin, which are not known to be teratogenic. Advise vitamin K1

prophylaxis against haemorrhagic disease of newborn if taking AEDs in last months of pregnancy.

51. Multiple Sclerosis 40 yo F (NICE Nov 2003)

DG: sxs? loss of colour discrimination, loss of central vision, double vision on looking outwards or down, difficulty driving at night (relapsing, remitting vs. progressive). IPS: empathy, ICE, family. Edx-CU. Denial. CMx: NICE 2004 Options: Ix – MRI, LP (CSF), VEP delayed. Prednisolone for optic neuritis, 1st line: β-interferon, glatiramer actetate. 2nd line tysabri. Refer neurology to make diagnosis. FU 1/52 with partner. Taxi. PILS.

Recommended Book by a GP with MS: George Jelinek: Taking Control of MS: Natural and Medical Therapies to Prevent its Progression.

BJGP Aug 2007 Prognostic factors for MS pain in 1° care a system review. C Mallen et al. 45 studies. Higher pain at baseline, longer pain duration, multiple site pain, previous pain episodes, anxiety +/ or depression, higher somatic perceptions +/ or distress, adverse coping strategies, low social support, older age, higher baseline disability and greater movement restriction associated with poor outcome.

WebMD: 28 April 2009: High doses of Vitamin D cut MS Relapses. High dose vitamin D reduces T cell activity (T lymphocytes order attacks on the myelin sheaths that surround and protect the brain cells.). 16% relapsed on 14000 IU vitamin D daily x 1 y vs. 40% on 1000 IU daily. J Burton Neurologist at Univ of Toronto.

The diagnosis is made clinically by a neurologist on the basis of evidence of CNS lesions scattered in space and time, on the basis of hx and exam. When the diagnosis remains in doubt, further ix should r/o alternate diagnosis or find evidence to support diagnosis (dissemination in space or time - MRI, dissemination in space – VEP studies, CSF analysis - for presence of oligloconal bands and compare with serum samples). The diagnosis of MS is clinical and an MRI scan should not be used in isolation to make the diagnosis. Sudden (w/n 12-48h) ↑ in neurological sxs or disability or develops new neurological sxs, a formal assessment should be made to determine the diagnosis.

Diagnosis of optic neuritis - acute, sometimes painful, ↓ or loss of vision in 1 eye. Common presenting symptom. Acute ↓ VA (+/- pain) refer ophthalmology. If the diagnosis is confirmed, the ophthalmologist should discuss the potential diagnosis and offer a further referral to neurologist. Tx: IV methylprednisolone 500 mg-1g daily x 3-5/7 OR high-dose oral methylprednisolone 500 mg - 2g daily for 3-5/7. Discuss risks/ benefits steroids.

Diagnosis of transverse myelitis - acute episode of weakness or paralysis of both legs, with sensory loss and loss of control of bowels and bladder is an emergency. Urgent ix: acute spinal cord dysfunction to r/o surgically treatable compressive lesion.

- Fatigue - aerobic exercise, energy conserving technique, rx amantadine 200 mg daily small evidence
- Bladder dysfunction - post-micturition residual bladder vol (U/S bladder), UTI, intermittent self-catheter, rx anticholinergic (oxybutynin or tolterodine) for urge incontinence. Check for ↑ in post-void residual volume. Nocturia rx desmopressin 100-400 ug orally or 10-40 ug intranasally nocte or for travel (not use > once in 24h period). Special continence service for assessment, pads,

convene drain (men), self-catheter, long-term urethral catheter, IV botox.

- UTI - cranberry juice, prophylactic antibiotics, assess by continence specialist for residual urine if 3 UTIs in 1yr
- Bowel - oral laxatives, suppositories, enemas
- Spasticity - neuro-physiotherapists, phys techniques to avoid contractures. Rx 1st-line baclofen or gabapentine. 2nd line tizanidine, diazepam, clonazepam or dantrolene. Splints, serial casting, intrathecal baclofen.
- Vision - $\downarrow$ read newspaper or watch tv due to poor control of eye movements, assess for low-vision equipment and adaptive technology, refer to specialist social services team, register partially sighted
- Neuropathic pain - carbamazepine, gabapentin or amitriptyline, refer
- Musculoskeletal pain - TENS or antidep
- Cognitive loss - formal cognitive neuro-psychological assessment with specialist clinical psychologist
- Emotionalism - cry or laugh with min provocation. Rx TCA or SSRI
- Depression - Do you feel depressed? Liaison psychiatrist if severe. Rx antidepressant or CBT
- Anxiety - Rx antidepressant or benzo
- Swallowing - PEG, swallowing technique using videofluoroscopy, short-term NG tube, chest physio
- Speech - dysarthria - speech and language therapist
- Sex- sildenafil 25-100 mg men
- Pressure ulcers - assess if using wheelchair

52. <u>BMJ 27 Jan 2007 Trigeminal neuralgia and its mx</u> L Bennetto et al. Neurosurgery dept, Bristol. Minority assoc w/ MS or N compression by tumour. Carbamazepine 1st line. Ablative surgical tx for drug failure, associated with facial sensory loss and high rate of pain recurrence. Microvascular decompression has risk of severe complications or death (0.4%) and a lower relapse rate. Aberrant loop of art or vein found compressing root entry zone of CNV in

80-90% at op. CNV is demyelinated next to compressing vessel. Intraop conduction better.

53. BJGP July 2006: Cluster HA in 1° care. D Kernick et al.

M 5x>F. 30s-40s yo. Unilateral HA, prefer to move about, so severe need to bang head against wall, 15 mins – 3 hours, associated with conjunctival injection, miosis, ptosis, lacrimation, rhinorrhoea, sweat, nasal block. Cluster period is 6-12 weeks. Preventative tx steroids 2-3 weeks tapering dose from 60 mg od for 5 d or methysergide. For long-term preventative tx, verapamil DOC. Lithium alternative.

54. Headache –migraine

BMJ 3 Feb 2007: Headaches G Fuller et al. British Assoc for the Study of Headaches (BASH 2004) Tx migraine Step 1 NSAIDs with buccal maxolon. Step 2 diclofenac IM +/- domperidone PR. Step 3 Sumatriptan 50 mg ↑ to 100 mg or suggest 20 mg nasal spray. C/I kids < 12 yo, IHD, CHD, CVD. Step 4 combo. Step 5 diclofenac with chlorpromazine IM emergency tx at home. Prophylaxis x 4-6/12. 1st line β-blockers. 2nd line sodium valproate and topiramate. 3rd line gabapentin. Then withdraw over 2-3 weeks.

BJGP Jan 2007: Patient pressure for referral for HA: a qualitative study of GPs' referral behaviour. M Morgan et al. HA = 1/3 of neurology outpatients with brain lesions rare.18 surgeries in S Thames, London. Factors: freq attenders, communication problems, time constraints, personal tolerance of uncertainty, differences in clinical confidence in identifying risks of CA, views of patients' 'right' to referral and therapeutic value of referral. Conclusion: Require further GP training to ↓GP referral to specialists as referral is often the outcome of patient pressure.

55. Irritable bowel disease - 20 yo F

<u>DG</u>: IBS Manning's criteria ($\geq$ 3 sxs for $\geq$ 3/12) (tenesmus, abdominal distension, relief with BO, looser motion with pain onset, more freq motion with pain, mucus), associated sxs (weight loss, fever, TAT, anaemia, loss of appetite). OE look for pallor, clubbing (IBD), LN, fever, oral ulcer, abdomen, chaperone PR rectal mass. <u>IPS</u>: ICE concerned as father has Crohn's. IADL, job, stress. EdxCU. Not CA and unlikely Crohns. <u>CMx:</u> Options - wait and see, diet (cut down caffeine, milk), food diary, mebeverine rx, $\uparrow$ bran, SSRI, CBT, stress counsellor. DTA. RGO. New Tegaserod 5-HT4 R partial agonist + a prokinetic in the gut? Refer for colonoscopy to make diagnosis as is diagnosis of exclusion. FU. PILS. If has red flags, refer urgently!

<u>GP Mag 8 Dec 2006</u> <u>Managing IBS</u> N Price. Diagnosis: At least 12 weeks (need not be consecutive) in preceding 12 months of abdo pain with 2 or 3 feat: relieved by defecation, onset assoc with change of stool freq or with change in stool form. Sxs supporting diagnosis: abnormal stool freq, abnormal stool form, abnormal stool passage, passage of mucus, bloating. Likely if < 45, longstanding and fluctuating, normal phys exam and no red flags (anorexia, weight loss, rectal bleed) fhx CA. Daily diary along with stressful events, food. $\uparrow$ bran. Tx: CBT, TCA, antispasm (mebeverine), antimuscarinics (hyoscine butylbromide), peppermint oil. Lactulose makes worse. Loperamide or codeine.

<u>BJGP Feb 2006 Sx interpretation and quality of life in patients with IBS.</u> B Bray et al. Edinburgh. Sx interpretation did not differ between IBS and non-IBS patients referred to hosp GI. Idea that most IBS patients are committed to a somatic explanation of sxs appears to be a myth. Normally a minority of IBS patients seek medical attention.

IBS IN ADULTS (NICE 08)

Any of these sxs for at least 6/12 (Abdo pain/discomfort, Bloating, Change in bowel habit) <u>Hx and exam by GP/ 1°care clinician</u> → IBS Positive Diagnosis Criteria Ixs in PC (FBC anaemia, ESR, CRP – Inflammatory bowel disease, EMA or TTG coeliac).

<u>Red Flag sxs</u> (Rectal bleeding, Unexplained unintentional weight loss, fhx bowel/ovarian CA, looser + frequent stools late onset > 60yo. Assess for anaemia, abdo, pelvic (if appropriate), rectal masses and inflammatory bowel disease. <u>Immediate referral to 2° care</u>.

<u>Diagnosis</u>: if the pt has abdo pain or discomfort that is either relieved by defaecation or associated with altered bowel frequency or stool form. This should be accompanied by at least 2 of the following 4 sxs: altered stool passage (straining, urgency, incomplete evacuation), abdominal bloating (F > M), distension, tension or hardness, sxs made worse by eating, passage of mucus.

The following tests are NOT necessary to confirm diagnosis who meet the IBS diagnostic criteria: U/S; rigid/flexible sigmoidoscopy; colonoscopy; barium enema; TFT; faecal ova and parasite test; faecal occult blood hydrogen breath test (for lactose intolerance and bacterial overgrowth). <u>Mx</u> base on nature and severity of sxs, individual or combination of meds, lifestyle advice, direct at main sx(s).

<u>Lifestyle</u>: Assess_diet:↓fibre intake; take soluble fibre; consider dietitian referral. Assess level of physical activity:_encourage ↑. Effective sx control. FU to evaluate response. Continuing sx profile. >12 months' duration, consider psychological interventions (hypnotherapy, psychological tx, CBT).

<u>Drug Tx</u>: single or combo: antispasmodics, anti-motility (titrate dose), laxatives, 2nd line TCA or SSRIs. Drink at least 8 cups of

fluid/day, water or other non-caffeinated, i.e. herbal teas. Restrict tea and coffee to 3 cups/day. ↓EtoH and fizzy drinks.

Limit intake of high-fibre food (i.e. wholemeal or high-fibre flour and breads, cereals high in bran, and whole grains, i.e. brown rice). ↓ intake of 'resistant starch' (starch that resists digestion in the small intestine and reaches the colon intact), which is often found in processed or re-cooked foods.

Limit fresh fruit to 3 portions/day (a portion approx 80 g). Patients with diarrhoea should avoid sorbitol, an artificial sweetener found in sugar-free sweets (including chewing gum) and drinks, and in some diabetic and slimming products. Patients with wind and bloating may find it helpful to eat oats (such as oat-based breakfast cereal or porridge) and linseeds (up to one tbs per day). Review the fibre intake of people with IBS, adjusting (↓) it while monitoring the effect on sxs.

Discourage from eating insoluble fibre (bran). If an ↑in dietary fibre is advised, it should be soluble fibre i.e. ispaghula powder or foods high in soluble fibre (oats). Advise how to adjust their doses of laxative or antimotility agent according to the clinical response. The dose should be titrated according to stool consistency, with the aim of achieving a soft, well formed stool (corresponding to Bristol Stool Form Scale type 4).

Consider TCAs as 2nd-line tx if laxatives, loperamide or antispasmodics have not helped. TCAs are primarily used for tx of depression but are only recommended here for their analgesic effect. Start at a low dose (5–10 mg equivalent of amitriptyline), which should be taken once nocte and reviewed regularly. The dose may be ↑, but does not usually need to exceed 30 mg. SSRIs should be considered for people with IBS only if TCAs have been shown to be ineffective.

Refractory IBS > 12 months - CBT, hypnotherapy and psychological tx– useful in helping to cope with their sxs, but it is unclear at what stage these should be given, including whether they should be 1st-line tx in 1° care.

56. Abdominal pain – acute cholecystitis
BMJ 11 Aug 2007 Clinical review: Gallstones G Sanders et al.10-15% Western pop. Fat, fertile, F, 40, Fhx. Problems: biliary colic/ cholecystitis, jaundice, ascending cholangitis, pancreatitis, Bouveret's syndrome (stone duodenal obstruction) and gallstone ileus (air in biliary tree), GB CA. Asymptomatic 1-4% watch and wait. Analgaesia (diclofenac and opioid rx), ursodeoxycholic acid prevents but not help once stones formed, percutaneous cholecystostomy drainage allows resolution of sepsis in patients at high surgical risk. Surgery: lap vs. open: Cochrane review: no difference in mortality, complications or op time. Shorter hospital stay and quicker convalescence. Early lap chole (< 7d of onset of biliary colic or cholecystitis) is safe and shortens hosp stay. Taunton study: 28.5% readmission with gallstone related complications for patients on WL for surgery after emergency admission with acute cholecystitis so advise op during emergency admission.

57. Acute Appendicitis
BMJ 9 Sept 2006 Clinical review: Acute appendicitis J Simpson. Most common surgical emergency 40,000/y, 10-20 yo. CT more sensitive than u/s. Lap appendicectomy ↑. ↓wound infections with periop antibiotics. Retrocaecal/ retrocolic 75%, subcoecal and pelvic 20%, pre and post-ileal 5%. FBC neut > 75%, leu 80-90%, CRP↑ but not always. Rovsig's sign (palpation of LIF causes pain in RIF), psoas stretch sign (pain with hip extension stretch psoas which overlies caecum so patient flexes hip), obturator sign. 8-10 cm long. After 36h rate of perforation 16-36% and ↑5% q 12h.

58. Diarrhoea – faecal impaction

59. BJGP Oct 2006: Predicting colorectal CA risk in patients with rectal bleeding R Robertson et al. Observational study. S England. 604 patients and 22 diagnosis with colon CA. Insufficient diagnosis value, component of a composite score. 2% with CA had presence of haemorrhoids. Significant predictors: age < 50 or ≥ 70 and blood mixed with stool.

60. BMJ 7 April 2007 Pre-endoscopy serological testing for celiac disease: evaluation of a clinical decision tool A Hopper et al. Pre-endoscopy testing for IgA tissue transglutaminase antibodies + duodenal biopsy of high risk patients = 100% sensitivity.

61. NICE Dyspepsia August 2004
Review meds: ca antagonists, nitrates, theophyllines, bisphosphonates, corticosteroids, NSAIDs.
Urgent specialist referral for endoscopy for patients of any age with dyspepsia and: Chronic GI bleed, Fe deficiency anaemia, progressive unintentional weight loss, epigastric mass, progressive difficulty swallowing, suspicious barium meal, persistent vomiting. Routine referral for endoscopy if > 55 and sxs persist despite H pylori testing and acid suppressing tx and when patients have ≥ 1 of: prior gastric ulcer or op, continuing need for NSAIDs, ↑ risk of gastric CA, anxiety re CA.

Start with PPI x 1 month OR test for and treat H pylori (carbon-13 urea breath test, stool antigen or lab-based serology). 2/52 washout period following PPI use before test for H pylori with breath test or a stool Ag test. Rx:7-day, bd course of full-dose PPI + (metronidazole 400 mg + clarithromycin 250 mg) or (amoxicillin 1g + clarithromycin 500 mg).
GORD - full-dose PPI for 1-2 mos. If sxs recur, offer a PPI at the lowest dose and limit repeat rxs
PUD - H pylori eradication if +. Stop NSAIDs. Offer full-dose PPI or H2RA tx for 2 mos.
Non-ulcer dyspepsia - initial tx for H pylori if +, symptomatic mx and periodic monitoring.

Guidelines from the British Society of Gastroenterology 2002

AGE FOR ENDOSCOPY – for new dyspepsia was 45 yo now 55 yo in line with national CA referral guidance. TEST AND TREAT – Treat patients < 55 yo with uncomplicated dyspepsia on basis of a + H pylori and not 'test and scope.' 13C UREA BREATH TESTS – the best test for ID and for confirmation of eradication of H Pylori is the 13C urea breath test. USE OF PPIs – continue to follow NICE guidance.

Testing for HP – serology is simple, widely available and has a high sensitivity but is less accurate than the urea breath test. Routine endoscopy for diagnosis of H Pylori is not recommended.

Endoscopy – for new onset of uncomplicated dyspepsia if > 55 (withhold antisecretory drugs for 4/52 before scope) and for patients with alarm symptoms if < 55. Alarm symptoms. Endoscopy is inappropriate for: DU which has responded symptomatically to tx; patients < 55; patients who have recently undergone satisfactory endoscopy for same sxs.

Alarm sxs: dysphagia and odynophagia, prior gastric surgery, epigastric mass, prior gastric ulcer, GI bleed, suspicious barium meal, persistent continuous vomiting, unexplained Fe deficiency anaemia, unintentional weight loss (≥ 3 kg)

Treatment for HP

- Duodenal ulcer +HP– 1 week triple tx with PPI (bd) or RBC (ranitidine bismuth citrate) +amoxycillin 500g –1g bd or metronidazole 400-500 mg bd + clarithromycin 500 mg bd; quadruple tx for 2nd line tx with PPI + bismuth 120 mg qds + metronidazole 400-500 mg tds + tetracycline 500 mg qds
- Duodenal ulcer no HP – cimetidine 800 mg nocte; refer GI if not NSAID-ulcer.

- Gastric ulcer + HP – Heliclear + antisecretory tx for 2/12. Long-term PPI or misoprostol if on NSAIDs.
- Gastric ulcer no HP- antisecretory therapy for 2/12. Stop NSAIDs. Give PPI if on NSAID.
- Oesophagitis – 4/52 of antacids, raft preparations (alginate) H2RA, or prokinetic agents (cisapride).

BJGP Oct 2006 GORD: a re-appraisal. Roger Jones KCL. Cardinal sxs heartburn and regurgitation. No gold standard diagnosis. 10-20% western prevalence. 15% of UK rxs are PPI. 'Montreal definition' 2004 in AmJGastroenterology. Non-pharmacological measures: stop smoking, obesity, EtOH, postprandial stoop, avoid late meals, raise head bed.

BMJ 6 Jan 2007 Dyspepsia and H pylori Rupal Shah. Endoscopy 30% normal, only 2% CA. Dyspepsia without alarm sxs, test and treat for H Pylori or give PPI is more economical than refer. Review patients on tx > 6/52 to step down or stop tx. Gastric ulcers on endoscopy need 4 weeks' tx with PPI and H pylori tx and then have repeat endoscopy (2% CA). For patients at high risk of PUD (old, h/o ulcers, or taking rx causing ulcers), who test + for H pylori, give eradication tx before starting regular tx with NSAIDs. Urea breath test most accurate ix (sensitivity 95%, specificity 95%) vs. serology (sensitivity 92.4%, specificity 91.9%). Need to stop antibiotics 4/52 before, PPI 2/52 before and H2R antagonist 1 day before test.

NICE 04: refer 2 week rule: chronic GIB, unintentional weight loss, difficulty swallowing, persistent vomiting, Fe deficiency anaemia, epigastric mass or suspicious findings on barium meal. Patients > 55 with 4-6/52 unexplained or recent onset even in absence of alarm. > 65yo on NSAID add PPI.

62. Dysphagia – pharyngeal pouch

63. Missed diagnosis of perforated ovarian cyst (based on real pt)
DG: 'Dr X did not examine me. He gave me fybogel for constipation and did a UPT when I said I was a virgin! I went to a female GP who examined me and arranged an urgent scan. I was in so much pain; I went to A+E. I was operated on that night and was told I had a 20 cm ovarian cyst which burst and was bleeding inside of me. I want this Dr X struck off! I almost died.' IPS: Sympathy. 'I'm so sorry. I understand your concerns. Thank you for bringing this to our attention.' Appreciate how serious this was for the pt and we will have emergency meeting – SEA so it doesn't happen again. Practice complaints procedure – last resort. Do not be defensive. CMx: Options: sxs mx – coc? FU with doctor concerned. 'Keep close eye on you in future.'

Recommended reading: BMJ: 9 Feb 2008 Vol 336 *Dealing with Complaints*, Judith Cave, Jane Dacre, UCL 'A survey of 1007 complainants found that their primary motivation was to prevent other patients experiencing the same adverse event in the future....'

Friel et al BMC Health Serv Res 2006; 6:106. *Patients 'expectations of fair complaint handling in hospitals':* admit a mistake when it has occurred, explain how the incident could have happened, offer an apology, show sympathy for what I went through, make an effort to recover our relationship.

64. TATT: BMJ 9 June 2007: Tiredness. G Moncrieff et al. ddx: depression, obesity, OSA, stress, hard work, poor sleep, caffeine withdrawal, anaemia, CA, renal disease, liver disease, heart failure, thyroid, diabetes, autoimmune. PE: pallor, LN, thyroid, ht F. Ix: fbc, ESR, lft, lytes, FBG, TFT, UA for protein & glucose. ?monospot, endomysial Ab, ANA Ag, CXR

65. Demands sick note (common GP request)
DG: Why? Hitler boss needs sick note. Job? Waitress. Home? DTA. IPS: do not get defensive. Keep language simple. Examine. EdiagnosisCU: URTI, SC2? CMx: self certify. Advise ACAS if it is a work-related problem. PILS practice policy on sick notes for employer, leaflet on minor ailments. Receptionist can help fill out if illiterate.

66. Demands orlistat (NICE Obesity 2006)
DG: Why? diet (snacking), sports, school/ job, family, FH obesity, DM, IHD. BMI. Waist circ. IPS: ICE. Low self-esteem. Bullying? Parent's expectation of orlistat. EdxCU. Causes of obesity. CMx: Options: Fruit + veg. Aerobic exercise. Orlistat only 18-75 yo and weight loss of ≥ 2.5 kg over 4 week by diet + exercise. Refer dietician, counsellor. Swimming. Contact school re bullying. Self help group. PILS. FU.

BJGP Sept 2006 Primary care support for tackling obesity: a qualitative study of the perceptions of obese patients. I Brown et al. 5 GPs in Sheffield. 28 patients. Patient's ambivalence, stigma and perceived lack of NHS resources.

BMJ 06; 332:505 Confusion among local professionals over best way to tackle childhood obesity. UK govt target to halt ↑ by 2010.

BMJ 05; 330 Overweight children ↓self esteem, perpetuates inactivity, overeating, adult obesity.

Watching tv major contributor for childhood obesity (Obesity Reviews 2005; 6:123-32).

Practitioner May 2006-08-05 What's new-tackling childhood obesity. D Haslam. Up to 75% become obese adults. Diet and lifestyle suggestions RCPCH/NOF - exercise, ↓ tv, walk to school, diet.

BMJ 13 May 06 Development of adiposity in adolescence: 5 yr longitudinal study of an ethnically and socioeconomically diverse sample of young people in Britain. J Wardle et al. Prevalence of overweight and obesity was high in London school students, with significant socioeconomic and ethnic inequalities. Few obese or overweight adolescents ↓ to a healthy weight. Persistent obesity is established before age 11.

NICE Obesity Dec 2006 Interventions for childhood overweight and obesity should address lifestyle w/n the family and social settings. BMI (adjust for age and gender) recommended. Interpret with caution as it is not a direct measure of adiposity. Consider specialist referral for kids. Significant comorbidity or complex needs (i.e., learning or educational difficulties). Prescribe orlistat as part of an overall plan for managing obesity in adults who meet 1 of the following: a BMI of ≥ 28.0 kg/m^2 with associated RFs or a BMI of ≥ 30.0 kg/m^2. Continued beyond 3/12 only if the patient has lost at least 5% of their initial body weight since starting drug treatment. The decision to use drug treatment for > 12 months (for weight maintenance) should be made after discussing benefits and limitations with the patient.

67. Fertility (NICE Feb 2004)

84% of couples conceive w/n 1 yr. 94% aged 35, 77% aged 38 after 3 yrs. Screen for chlamydia before uterine instrumentation. F < 1-2 Us once or twice a week, M < 3-4 Us/d. Offer folic acid supplement before conception and up to 12/40. 5 mg a day if on antiepileptic rx. Offer rubella susceptibility screening and vaccination. Infertility: failure to conceive after regular UPSI for 2 years in the absence of reproductive pathology. If unable to conceive after a year, offer semen analysis +/or assess ovulation. Couples seen together.

Semen Analysis: volume $\geq$ 2 ml; liquefaction time: within 60 mins; pH $\geq$ 7.2; sperm concentration 20 million per ml; total sperm number: 40 million per ejaculate; 50% motility or more; vitality 75%; wbc < 1 million per ml; morphology: 15% or 30%. Screening for antisperm antibodies should NOT be offered as there is no effective treatment to improve fertility. If the result of the first semen is abnormal, repeat test 3 months after initial analysis to allow time for cycle of spermatozoa formation to be completed. If azoospermia, then repeat ASAP.

Assessing ovulation

- Offer serum progesterone in the mid-luteal phase (day 21 of a 28-day cycle) to confirm ovulation in F with regular menses and > 1 year infertility. If prolonged menses, offer test later in cycle day 28 of 35-day cycle and repeat weekly until the next menstrual cycle starts.
- Use of basal body temp is NOT recommended. Routine TFTs are NOT recommended.
- Irregular menstrual cycles - offer serum FSH and LH tests.
- Measure prolactin ONLY if have an ovulatory disorder, galactorrhoea or a pituitary tumour.
- Tests of ovarian reserve have limited sensitivity and specif. Using Inhibin B is uncertain in assessing reserve.

- Endometrial biopsy should NOT be offered to evaluate the luteal phase.

Assessing tubal damage
- Offer HSG to screen for tubal occlusion if not known to have co-morbidities (PID, previous ectopic or endometriosis). Hysterosalpingo-contrast-ultrasonography is an effective alternative.
- If known to have co-morbidities, offer laparoscopy and dye. Assessing uterine abnormalities - NOT offer hysteroscopy on its own as part of initial Ix.
Postcoital testing of cervical mucus is NOT recommended.

Medical and surgical mx of male factor fertility problems
- Offer gonadotrophin drugs to men with hypogonadotrophic hypogonadism.
- Idiopathic semen abnormalities should NOT be offered anti-oestrogens, gonadotrophins, androgens, bromocriptine or kinin-enhancing drugs.
- Effectiveness of systemic steroids for antisperm antibodies is uncertain.
- Men with leukocytes in their semen should NOT be offered antibiotics without identified infection.
- Surgical correction of epididymal blockage for obstructive azoospermia.
- Surgery for varicocoeles should NOT be offered, as it does not improve pregnancy rates.

Ovulation Induction: 21% of F fertility probs. WHO Group I: hypothalamic pit failure (hypothalamic amenorrhoea or hypogonadotrophic hypogonadism); Group II: hypothalamic pit dysfunction (PCO); Group III: ovarian failure.
Antioestrogens: PCO should be offered clomifene citrate (or tamoxifen) as 1st line for up to 12 months to induce ovulation and told of risk of multiple pregnancies. Offer U/S monitoring during 1st cycle. Metformin - is NOT currently licensed for the tx of

ovulatory d/o. Anovulatory F with PCO who have not responded to clomifene and have a BMI > 25 should be offered metformin combined with clomifene. Warned of side-effects: N/V, GI.
Laparoscopic ovarian drilling - PCO and not respond to clomifene.

Gonadotrophins: PCO and not ovulate on clomifene, may offer human menopausal gonadotrophin, urinary FSH and recombinant FSH. If on clomifene and not pregnant after 6 months, offer clomifene citrate - stimulates intra-uterine insemination. Used following pituitary-down regulation as part of IVF. PCO and on gonadotrophins should NOT be offered gonadotrophin-releasing hormone agonist as associated with ↑ risk of ovarian hyperstimulation. GnRH agonists in long protocols are routinely used during IVF.
Use of growth hormones as an adjunct to ovulation induction therapy is NOT recommended.
Pulsatile Gn releasing hormone should be offered to Group I ovulation disorders.

Dopamine agonists - bromocriptine for hyperprolactinaemia.
For proximal tubal obstruction, offer selective salpingography + tubal catheterisation or hysteroscopic tubal cannulation.
Hysteroscopic adhesiolysis for intra-uterine adhesions.

Medical and surgical mx of Endometriosis
- Medical tx (ovarian suppression) should NOT be offered.
- For minimum or mild endometriosis, offer surgical ablation or resection of endometriosis + lap adhesiolysis.
- Ovarian endometriomas should be offered laparoscopic cystectomy.
- Moderate or severe endometriosis offer surgical tx. Post-op medical tx does not improve pregnancy rates.

Intra-uterine insemination
- Mild male factor, unexplained fertility, or minimal to mild endometriosis should be offered up to 6 cycles.

- Ovarian stimulation should NOT be offered because of risk of multiple pregnancy.
- Single rather than double insemination should be offered.
- Fallopian sperm perfusion for insemination should be offered for unexplained fertility

Offer salpingectomy for women with hydrosalpinges.

IVF: Chances of live birth per IVF tx cycle: > 20% 23-35yoF; 15% 36-38yo; 10% 39 yo; 6% ≥ 40 yo.
- Screen for HIV, hep B and C; Adversely affect success rate of IVF: EtOH > 1 U /day, tobacco, caffeine, BMI > 30
- Couples in which the woman is 23-39 yo at time of tx and who have an identified cause of infertility (azoospermia or bilateral tubal occlusion) or who have infertility of min 3 years offer up to 3 stimulated cycles of IVF.
- Human menopausal gonadotrophin, urinary FSH and recombinant FSH are equally effective in achieving a live birth when used following pituitary down-regulation as part of IVF.
- Inform couples of chance of multiple pregnancy after IVF depends on the no of embryos transferred per cycle of tx. No more than 2 embryos should be transferred during any 1 cycle of IVF.
- Embryos not transferred during a stimulated IVF tx may be suitable for freezing. If 2 or more embryos are frozen then they should be transferred before the next stimulated tx cycle to minimise ovulation induction and egg collection, both of which carry risks for the woman and use of more resources.
- Clomifene citrate-stimulated and gonadotrophin-stimulation IVF have higher pregnancy rates than natural cycle

Intracytoplasmic sperm injection - severe deficits in quality, obstructive azoospermia, non-obstructive azoospermia; genetic counselling, karyotype testing. Testing for Y chromosome microdeletions NOT routine ix.

<u>Donor insemination</u> – obstructive azoospermia, non-obstructive azoospermia, HIV male, severe rhesus isoimmunisation, severe deficit in semen quality, risk of genetic d/o of offspring

<u>Oocyte donation</u> - premature ovarian failure, Turners, bilateral oophorectomy, ovarian failure after chemo or radiotx, certain cases of IVF fails. Application of cryopreservation in cancer tx

68. <u>Long-Acting Reversible Contraception (NICE Oct 05)</u>
LARC-copper IUD, progestogen-only (IUS, injectables, subdermal implants) or combined vaginal rings. More cost effective than coc even at 1 yr of use. IUDs, IUS and implants more cost effective than injectables.

Counselling: efficacy, duration of use, risk and possible S/Es, non-contraceptive benefits, procedure for initiation and removal/ discontinuation, when to seek help while using the method

<u>Contraceptive prescribing</u> - med hx, fhx, menstrual, contraceptive and sexual hx. Exclude pregnancy. Supply interim method of contraception at 1st appointment if required; obtain informed consent, refer VTE to contraception specialist. Promote safe sex. Assess risk for STIs and advise testing when appropriate. Info on local STI screening. Fraser guidelines for < 16 yos. Contraception in terms of needs of patient rather than relieving anxieties of carers or relatives. If learning disability, carers and parties involved should establish a care plan.

IUD and IUS only fitted by trained person with continuing experience of inserting min. one IUD or IUS a month.

<u>Copper IUD</u> IUDs prevent fertilisation and inhibit implantation. The licensed duration of use containing 380 mm^2 copper ranges from 5-10 yrs. Pregnancy rate is < 20 in 1000 over 5 yrs. No evidence of delay in return of fertility following removal or expulsion. Heavier bleeding +/or dysmenorrhoea are likely. Up to 50% stop using IUDs w/n 5 yrs due to unacceptable PVB and pain.

Does not affect weight. Risk of uterine perforation at time of insertion is < 1 in 1000. Risk of PID is < 1 in 100. Risk of expulsion is < 1 in 20 in 5 yrs. Overall risk of ectopic is 1 in 1000 in 5 yrs. If pregnant with IUD in situ, risk of ectopic is 1 in 20. ≥40 yo may retain device until no longer require contraception. Test for chlamydia, N gonorrhoea, and any STIs in women who request before IUD insertion. If not possible, then give prophylactic antibiotics before IUD insertion in those at ↑ risk of STIs.

May be used in adolescents but consider STI risk. Not C/I in nullips of any age. Not C/I in diabetics. Safe to used when breastfeeding and for HIV + or with AIDS. May be inserted at any time during the cycle, immediately after 1st or 2nd trimester abortion or from 4 weeks postpartum.

Emergency drugs (AEDs) should be available at time of insertion in epileptics as ↑ risk of seizure at time of cervical dilation. Check threads, light bleed for few days and pain for few hrs, see sooner if sxs of perforation or infection. F/U after 1st menses or 3-6 weeks after insertion to r/o infection, perforation or expulsion. Tx heavier periods with NSAIDs and tranexamic acid or if unacceptable, change to LNG-IUS. Presence of Actinomyces organisms on smear with IUD requires r/o PID. Routine removal not indicated. Remove IUD < 12 weeks gestation if have IUP.

IUS Acts predominantly by preventing implantation and sometimes fertilisation. Pregnancy rate is < 10 in 1000 over 5 yrs. Licensed use for 5 yrs duration. No delay in return of fertility. Irregular bleed and spot for 1st 6 mos. Oligo or amenorrhoea by the end of the 1st yr. Up to 60% stop IUS w/n 5 yrs due to unacceptable PVB and pain. No weight gain. ↑ acne as a result of absorption of progestogen.

Risk of uterine perforation at insertion < 1 in 1000. Risk of PID < 1 in 100. Expulsion < 1in 20 in 5 yrs. Risk of ectopic < than no contraception. Overall ectopic risk is 1 in 1000 in 5 yrs. Ectopic 1 in 20 if pregnant. $\geq$45 yo may retain IUS until they no longer require contraception. STI testing before insertion.

May be used in adolescents but consider STI risk; not C/I in nullips and IS safe in breastfeeding. Is safe to use if oestrogen is C/I. Safe for women with HIV or AIDs. Not C/I in diabetics. Insert at any time during cycle (but if amenorrhoeic or it has been > 5 days since menstrual bleeding started, additional barrier contraception should be used for the 1st 7 days after insertion); immediately after 1st or 2nd trim abortion or any time thereafter; from 4 weeks postpartum. Emergency drugs including AED to be made available at time of IUS insertion in epileptics. F/U after 1st menses or 3-6 weeks after insertion.

Progestogen-only injectables Act by preventing ovulation. Pregnancy rate is < 4 in 1000 over 2 yrs and the rate of DMPA is < NET-EN. DMPA (depot) q 12/52 and NET-EN (norethisterone enantate) q 8/52. Delay of up to 1 yr in return of fertility. If she stops, she should use a different contraception method immediately, even if amenorrhoea persists.

Amenorrhoea is more likely with DMPA than NET-EN, as time goes by and is not harmful. DMPA may be associated with ↑ of up to 2-3 kg in weight over 1 yr. Is NOT associated with acne, depression or HA's. DMPA is associated with a small loss of BMD, which largely recovers when stopped. No evidence of ↑ in fracture. May be given in adolescents or > 40 yo if other methods are not acceptable.

BMI > 30 can safely use DMPA and NET-EN and breastfeeding is safe. May be used by women who have migraines +/- aura. Medically safe if oestrogen is C/I. May be associated with ↓ in freq of seizures in epileptics. No evidence of ↑ risk of STI or HIV. Safe if HIV + or AIDS. May use if taking liver enzyme-inducing meds and the dose interval does not have to be reduced.IM into gluteal or deltoid or lateral thigh. Start up to and including the 5th day of the cycle without the need for additional cover. At any other time, need to use additional barrier for 1st 7 days after injection; may be used immediately after 1st or 2nd trim abortion or any time thereafter or at any time postpartum. May give if up to 2 weeks late for repeat DMPA without need for additional contraception. Persistent PVB associated with DMPA tx with mefenamic acid or ethinyestradiol. Review if wish to continue beyond 2 yrs. No evidence of congenital malformations if pregnancy occurs on DMPA.

<u>Progestogen-only subdermal implants (implanon)</u> Prevents ovulation. Pregnancy rate < 1 in 1000 over 3 yrs. Use for 3 yrs. No evidence of delay in return of fertility. Bleeding patterns likely to change. 20% have no bleed, while 50% have infrequent, frequent or prolonged PVB. Likely to remain irregular over time.

Dysmenorrhoea may be ↓ during the use of implanon. Up to 43% stop using implanon w/n 3 yrs due to 33% irregular bleed and < 10% hormonal problems. Not associated with weight change, mood, libido or HA's. May be associated with acne.> 70 kg can use. Safe in breastfeeding. Not C/I in diabetes. No ↑ risk of STI or HIV. Safe for HIV + or AIDs.

May be used in patients with migraine +/- aura. Medically safe if oestrogen is C/I; no evidence of effect on BMD. Implanon is NOT recommended for women taking liver enzyme-inducing drugs. Insertion: at any time (but if amenorrhoeic or > 5days since menses started, additional barrier for 1st 7 days; immediately after abortion and at any time post-partum. No routine F/U. Tx irregular bleed with mefenamic acid, ethinyestradiol or mifepristone. No teratogenic effect. If implanon cannot be palpated (due to deep or failed insertion or migration), localise by USG before remove.

69. <u>40 yo F Menorrhagia and dysmenorrhoea</u>
   <u>DG:</u> Amount, flooding, clots, IMB, PCB, assoc sx, LMP, ICE, IADL, R/O anaemia, endometriosis, thyroid. Contraception? OE

BP, goitre, pelvic. IPS: embarrassment over flooding. Concerns re CA, abnormal bleeding? EdxCU:DUB. DDx: fibroid, polyp. CMx: RCOG Options - nil, antifibrinolytic (tranexamic acid 1g tds x 3/7 at the start of each menses, mefenamic acid 500 mg tds prn, norethisterone 5 mg tds (for flooding or delay menses), coc, FeSO4 (FBC, ferritin), Mirena. Ix: TV scan, refer if PMB, postcoital / IMB or fibroids. Start pill? PN health ed. FU. RGO. DTA. Smear. PILS. If fibroid, med mx (tranexamic + mefenamic acid) or refer for SC zoladex into ant abdo wall q 28 days x 3/12 prior to myomectomy.

HEAVY MENSTRUAL BLEEDING (NICE Jan 2007)

Hx taking, exam and ixs If the hx suggests HMB without structural or histological abnormality, drug tx can be started without carrying out an exam or other ix's at initial consultation in 1° care, unless the tx chosen is LNG IUS. If the hx suggests HMB with structural or histological abnormality, with sxs i.e. IMB or postcoital bleed, pelvic pain +/or pressure sxs, perform an exam and/or other ix's (scan). Carry out exam before all: LNG-IUS fittings, ix's for structural and for histological abnormalities. Women with fibroids that are palpable abdominally or who have intracavity fibroids and/or whose uterine length as measured at scan or hysteroscopy is > 12 cm refer immediately to a specialist.

Ix's: FBC should be carried out on all. This should be done in parallel with any HMB tx offered. Consider test for coagulation disorders (von Willebrand's disease) in women who have had HMB since menarche and have personal or family history suggesting a coagulation disorder. A serum ferritin test and hormone testing should NOT routinely be carried out on women with HMB.TFT's should be carried out ONLY when other signs/ sxs of thyroid disease are present. If appropriate, a biopsy should be taken to exclude endometrial CA or atypical hyperplasia. Indications for a biopsy: persistent IMB, and in women aged ≥ 45 yo, treatment failure or ineffective treatment.

Imaging should be undertaken in the following circumstances: The uterus is palpable abdominally. Vaginal exam reveals a pelvic mass of uncertain origin. Drug therapy fails. Ultrasound is the 1st-line diagnostic tool for identifying structural abnormalities.

Hysteroscopy should be used as a diagnostic tool only when U/S results are inconclusive, i.e., to determine the exact location of a fibroid or the exact nature of the abnormality. If imaging shows the presence of uterine fibroids then appropriate tx should be planned based on size, number and location of the fibroids. Dilatation and curettage alone should NOT be used as a diagnostic tool. Women should be made aware of the impact on fertility that any planned surgery or uterine artery embolisation (UAE) may have, and if a potential tx (hysterectomy or ablation) involves the loss of fertility.

Drug rx: If hx and ix's indicate that pharmaceutical tx is appropriate and either hormonal or non-hormonal txs are acceptable, txs should be considered in the following order: LNG-IUS provided long-term (at least 12-mos); tranexamic acid or NSAIDs or COCs; norethisterone 15 mg od from days 5 to 26 of menses, or injected long acting progestogens. If hormonal tx's are not acceptable to the pt, then use either tranexamic acid or NSAIDs. When HMB coexists with dysmenorrhoea, NSAIDs should be preferred to tranexamic acid.

Ongoing use of NSAIDs and/or tranexamic acid is recommended for as long as it is found to be beneficial. Use of NSAIDs +/or tranexamic acid should be stopped if it does not improve sxs w/n 3 menstrual cycles. Use of a GnRH analogue could be considered prior to surgery or when all other tx options for uterine fibroids, including surgery or uterine artery embolisation, are contraindicated. If this tx is to be used for > 6 months or if adverse effects are experienced then recommend HRT 'add back' therapy. Oral progestogens given during the luteal phase should NOT be used for the tx of HMB. In women with HMB alone, with uterus < than a 10-week pregnancy, <u>endometrial</u> <u>ablation</u> should be considered preferable to hysterectomy. Women must be advised to avoid subsequent pregnancy and on the need to use effective contraception, if required, after endometrial ablation.

Endometrial ablation should be considered in women with HMB who have a normal uterus and also those with small uterine fibroids (< 3 cm in diameter). UAE, myomectomy or hysterectomy should be considered in cases of HMB where large fibroids (> 3 cm in diameter). Women should be informed that UAE or myomectomy may potentially allow them to retain their fertility. Pretx before hysterectomy and myomectomy with a GnRH analogue for 3/12 to 4/12 should be considered where uterine fibroids are causing an enlarged, distorted uterus.

If a woman is being treated with GnRH analogue and UAE is then planned, theGnRH analogue should be stopped as soon as UAE has been scheduled. <u>Hysterectomy</u> - 1st line vaginal; 2nd line abdominal.

70. <u>Vaginal discharge and dysuria – fear of diabetes –22 yo pregnant F</u>

<u>DG</u>: On coc for 3/12. Used OTC canesten cream but not working. Newly married. Job? <u>IPS</u>: ICE. Concern: Diabetes. +FHx. OE chaperone PV thrush + UA. EdiagnosisCU –Thrush and UTI. Need for pessary. Prevention advice. <u>CMx:</u> Options: HVS, blood test for FBG (+Fhx), rx amoxicillin for UTI in pregnancy, canesten pessary 500 mg vs. pessary x 3 200 mg + cream vs. flucan. Thrush prevention: avoid nylon underwear, tight trousers. UTI prevention: wipe front to back, empty bladder, + fluids, void pre/post sex. Safe sex, current pap. PILS. FU for HVS, MSU and FBG results. + FBG = be prepared to explain. Diabetes UK. Self-monitor. Hypos. Refer to diabetic clinic, DECS (diabetes eyes complication service), chiropody, antenatal clinical. Medic-alert bracelet.

71. <u>Gonorrhoea – 20 yo F 6/52 pregnant</u>

<u>DG</u>: Yellow-green PV discharge, 1 night stand, PH of STIs, inform partner, contact tracing. OE - T, abdo, chaperone, PV, swabs? <u>IPS</u>: Anonymity, confidentiality. ICE of STD. EdxCU-gonorrhoea vs. trich vag (depends on results). <u>CMx:</u> Options: Full STI screen GUM (GC: oral, endocervix, urethra and anus), test of cure. Safe sex condom. Rx Amox 3 g (pregnant) or Cipro 500 mg STAT (not pregnant). PN sex health ed. Gen contraception advice.

72. <u>15 yo F requests emergency contraception</u>

<u>DG:</u> Why? UPSI when? Condom or pill failure (EC)? CU pill and STIs. BF's age? Sex Offences Act 2003 consent SI 16. ≥18 yo-schedule 1 sex offence- breach to child protection services/ police. School, friends, LMP. UPT if sxs. OE: BP. Weight. <u>IPS</u>: Consensual? Confidential. Relationship with mother, bf? Family dynamics. EdiagnosisCU. Assess Frazer competency 'UPSIS' - understand facts and weigh pros/cons of pill. C/I to coc (DVT, FH breast ca, focal migraine). L-2 95% effective w/n 24h, 85% at 48h,58% at 72h, IUCD 99%. EC indications (miss >/= 4 coc in midpack, 2 coc in pill-free interval or 1st week or 2 pop). <u>CMx:</u> Options: STI screen (UPSI). Depo is not 1st-line for teens – osteoporosis; condoms; coc. CU. PN advice. FU if miss period. FPA, Youth Health Centre. Advise talk to parents. Long-term contraception. <u>Sources: FFP April 2000 Guidance sheet on Emergency Contraception, RCOG and BMA guidance on confidentiality and people under 16.</u>

73. <u>Preconception Counselling for Down's 40 yo F (ANTENATAL CARE – NICE Oct 2003)</u>

<u>DG:</u> Why now? On pill? TOP? <u>IPS:</u> concern re Down's. Husband supportive? EdiagnosisCU. Risk of Down's ↑ with age - 1/100 at 40, 25% at 45. Advise on diet, work stress, cat litter. <u>CMx:</u> Options: Screening: 11-14 weeks: Nuchal translucency scan (NT),

the combined test (NT, hCG and PAPP-A); Bart's triple blood test (14-20 weeks ↓AFP, unconjugated oestradiol (uE3), βHCG detects 50%), the quadruple test (hCG, AFP, uE3, inhibin A) from 11-14 weeks and 14-20 weeks - the integrated test (NT, PAPP-A, +hCG, AFP, uE3, inhibin A), the serum integrated test (PAPP-A, +hCG, AFP, uE3, inhibin A). Diagnostic: CVS 9 weeks, amnio 16 weeks. Ovulate 2 weeks before next period is due. Rx folic acid 400 mcg od. DTA. Vaccine - rubella, chickenpox. Antenatal blood tests. FU with husband. Pap.

74. 50 yo F requests HRT

DG: why? Vasomotor (hot flush, night sweats, HA, urinary sxs, vag dryness, forgetfulness) OE BP, weight. FSH> 30. IPS: ICE. Impact on family. Concern re hormone, menopause. EdxCU: Pros/cons.HRT only for s-t sx control. Breast CA RR 1.2. 3-5x↑ risk endometrial CA. 2x ↑ risk of DVT in 1st year. Risk of gallstones. Not for osteo or cardioprotection. S-t side effects of oestrogen: breast tenderness, HA, leg cramps and for prog: irregular bleed, bloating and weight gain. CMx: Options: HRT (creams, nasal spray, implants, transdermal patches, pessaries or tablets. Types: sequential or continuous combined, unopposed oestrogen (post hysterectomy), progestogen only, raloxifene (SERM licensed for osteo), tibolone (protects against osteo), British Menopause Society. LF. Age < 54 + bleeding = premique or prempak C (monthly bleed HRT). Age >/=54 + min 1yr postmen =nonbleed prep=premique. Premarin if no uterus. FU 2/52 to decide. HP: Pap/mammo. Osteo: Sunlight, diet, weight-bear exercises, calcichew forte. PILS. BMJ 7 April 2007:

Managing the menopause H Roberts. Between 40-58. HRT ↑RR CVA 1.4. Most start with 0.3 mg conjugated equine oestrogen, 0.5 mg-1 mg 17b-estradiol or estradiol valerate) to relieve flushes. Add progestogen if no uterus. May ↑ after few weeks. Combined continuous oestrogen and progesterone postmenopause > 1 yr, irregular spot then no bleed after 1st yr. < 1y: sequential oestrogen on d1 of period then progestogen d14 x 10-14d and light period bleed. TV US only if bleed after 1 yr no periods or IMB before HRT. Hot flushes/ night sweats: HRT or tibolone. Vag atrophy, recurrent UTIs: low dose topical oestrogen (best evidence), intravaginal oestrogen, tibolone or vaginal moisturiser. 40-50% stop HRT after 1y and 65-76% stop within 2 years.

75. Requests Tubal Ligation

DG: why now? Children, ages? Current contraception? What is your understanding? Considered vasectomy?
IPS: ICE. Financial or relationship problems? Regret? If children die or you divorce? ICE? Explain op. 1 in 200 failure, 1 in 10000 death from GA vs. vasectomy local (need 2 negative SA at 10 and 14 weeks).
CMx: Options: Implanon 3 years, depo 12 weeks, IUS/IUD. Discuss with spouse. Decide next visit. PILS.

76. 36 week check ANTENATAL CARE - NICE 2003, 2008 DG: sxs? concerns? OE: UA, BP, check position of baby (breech?), fundal height, Doppler. www.babyfriendly.org.uk care of new baby, baby blues, breastfeeding info.

NICE ANTENATAL CARE Mar 2008

Booking appt (ideally 10 weeks) Give info: how the baby develops during pregnancy; nutrition and diet, vitamin D supplementation; exercise, i.e. pelvic floor exercises; risks and benefits of the antenatal screening tests; pregnancy care pathway ; place of birth; breastfeeding workshops; participant-led antenatal classes; maternity benefits.

10 Weeks: check blood group and rhesus D status (ideally before 10 weeks), offer screening for haemoglobinopathies, anaemia, red-cell alloantibodies, hepatitis B virus, HIV, rubella susceptibility and syphilis (ideally before 10 weeks), UA to screen for

asymptomatic bacteriuria and proteinuria, inform pregnant women < 25 years about the high prevalence of chlamydia infection at their age, and give details of their local National Chlamydia Screening Programme (www.chlamydiascreening.nhs.uk), offering screening for Down's syndrome ('combined test' at 11 weeks to 13 weeks 6 days , serum screening test (triple or quadruple) at 15 to 20 weeks).

- offer early scan for gestational age assessment (crown–rump measurement 10 weeks to 13 weeks 6 days, head circumference if crown–rump length is > 84 mms) and for structural anomalies (between 18 and 20 weeks 6 days).
- measure height, weight and calculate BMI; measure BP and test urine for protein
- offer screening for gestational diabetes and pre-eclampsia using risk factors
- identify those with genital mutilation , ask about any past or present severe mental illness or psychiatric tx
- ask about mood to identify possible depression; ask about the woman's job to identify potential risks.
- Risk of listeriosis: Drink pasteurised or UHT milk; avoid mould-ripened cheese (camembert, brie and blue-veined), pate, undercooked meals
- Risk of salmonella: avoid raw or undercooked eggs, mayo, raw meat especially poultry
- Risk of toxo: avoid cat faeces in litter or soil, wear gloves when gardening, cook raw meats, wash all fruits and vegetables, wash hands before preparing food
- Advice on diet, lifestyle (no EtOH, tobacco, cannabis), long haul (DVT), maternity benefits, screening tests
- Screen for pre-eclampsia: nullip, > 40 yo, FH, prior hx, BMI > 35, multiple pregnancies or pre-existing vascular disease (DM, HTN), symptoms: headache, blurred vision, flashers, pain below ribs, sudden swelling of face, hands or feet.

16 weeks: Review results of all screening tests. Investigate Hb < 11 g/100 ml (consider iron), BP, test urine for protein. Give info of the routine anomaly scan.

18 to 20 weeks, in outpatients, scan for the detection of structural anomalies. If placenta is found to extend across the internal cervical os at this time, offer another scan at 32 weeks.

28 weeks (nullips) measure and plot symphysis–fundal height; BP and test urine for protein; give info.

28 weeks (all) give info, offer a 2nd screening for anaemia and atypical red-cell alloantibodies. Investigate hb < 10.5 g/100 ml and consider iron; offer anti-D prophylaxis to rhesus-negative women; BP and test urine for protein; measure and plot symphysis–fundal height.

31 weeks (nullips) BP, test urine for protein; measure and plot symphysis–fundal height; give info; review, discuss and record the results of screening tests undertaken at 28 weeks.

34 weeks (all) Give info on preparation for labour and birth, including info about coping with pain in labour and the birth plan; recognition of active labour. Offer a 2nd dose of anti-D to rhesus-negative women BP and test urine for protein; measure and plot symphysis–fundal height. Review, discuss and record the results of screening tests undertaken at 28 weeks

36 weeks (all) Topics to cover: breastfeeding info, including technique and good mx practices that would help a F succeed, i.e. detailed in the UNICEF 'Baby Friendly Initiative' (www.babyfriendly.org.uk) care of the new baby, vitamin K prophylaxis and newborn screening tests, postnatal self-care, awareness of 'baby blues' and postnatal depression. BP and test urine for protein; measure and plot symphysis–fundal height; check

position of baby; babies in the breech presentation, offer external cephalic version (ECV).

38 weeks will allow for: BP and urine testing for protein; measure and plot of symphysis–fundal height, info giving, including options for mx of prolonged pregnancy.

40 weeks (nullips): BP and test urine for proteinuria; measure and plot symphysis–fundal height, give info, options for prolonged pregnancy, verbal info supported by antenatal classes and written information.

For women who have not given birth by 41 weeks: offer info, a membrane sweep, induction of labour. BP and urine tested for proteinuria; symphysis–fundal ht should be measured and plotted

77. Breast lump/ pain 28 yo F (NICE 2005 Cancer)

DG: R/O CA, FH CA, tobacco, coc, PH breast problems, ICE, Job, Home. OE offer chaperone, breast model, LNs. EdxCU. Not CA? Options: breast pain cyclical – support bra, wait see, stop coc, gamolenic acid / noncyclical – mefenamic acid, danazol, bromocriptine/ refer if local pain or persists > 6/52/breast lump refer. Mammo 50-64.

GP Mag 13 October 2006 Breast CA screening I Locke. #1CA in F UK. 1in 9. TP53 mutation. BRCA1 and BRCA2 ↑ breast and ovarian CA and diagnosis < 40 yo. ↑Icelandic and Ashken Jews. Mod risk mammo 40-49. NHSBSP mammo 50. MARIBIS study 2005 breast MRI more sensitive than mammo for detect breast CA.

Discrete, hard lump with fixation, +/- skin tethering.

Age < 30: benign lumps (fibroadenoma) or breast pain + no palpable abnormality, consider non-urgent.

Age < 30: lump that enlarges, or is fixed and hard, or reason for concern i.e. FH.

Age ≥ 30 with a discrete mass persisting after next period or presenting after menopause Any one of: spontaneous unilateral bloody nipple, unilateral eczematous skin or nipple change not responding to tx; nipple distortion of recent onset; previous histologically confirmed breast CA + lump or suspicious sxs.

Men ≥ 50 with unilateral, firm subareolar mass +/- nipple distortion or associated skin changes.

BMJ 18 Aug 2007: Referral patterns, CA diagnoses and waiting times after intro of 2 week wait rule for breast CA: a prospective study S Potter et al. Bristol '99-'05. Over 7 yrs, # of CAs detected (↓12.8% to 7.7%) in 2 week wait pop↓ and unacceptable % now referred via routine route (↑2.5% to 5.3%). Need to review the 2 week wait rule urgently! 27% of CA are referred in the non-urgent group! 2 week: patient of any age: discrete hard lump with fixation, unilateral eczema change, nipple distortion, spontaneous unilateral bloody discharge, skin distort, new lump in patient with previous diagnosis breast CA. F > 30 with discrete lump after period/ menopause; fixed, hard or enlarging lump. M > 50 with unilateral firm subareolar mass.

78. 1° dysmenorrhoea

DG: 14 yo F painful periods. Ask sxs of PMT. Ask psychological (sexual abuse) vs. organic (sexually active)? EdxCU. CMx: mefenamic acid (PG synthetase inhibitor) 500 mg tds prn or coc.

BMJ 13 May 2006: Diagnosis and Mx of dysmenorrhoea. M Proctor et al. Common and under diagnosis and under treatment. Simple analgaesics and NSAIDs effective in up to 70%. Oral contraception/ IUS considered for those who wish to avoid pregnancy. Alternative tx-heat, thiamine, Mg, vitamin E effective. 1^0 6-13 mos after menarche. Refer if coc/ NSAIDs no help. RF - stress, obesity, EtOH, tobacco, younger age menarche, depression.

79. Pre-menstrual tension

DG: PMT sxs (bloating, anxiety, breast tenderness, depression, HA, hostility. Menstrual hx, contraception, pap, caffeine? IPS: ICE? Home? Job? Stress? EdxCU: PMT. CMx: Options - wait see, diary, relax, exercise, cut out refined sugar, caffeine, add small frequent CHO, nsaids. RCOG guidance: 1st line: 3rd gen tricycle or bicycle coc (yasmin or cilest), 2nd line: oestradiol patches 100 ug + oral progestogen Duphaston 10 mg or Mirena or citalopram 10-25 mg on days 15-28. Rx. RGO. FU. DTA. SSRIs UK unlicensed for PMT now. Used to be available for disabling PMT (DRCOG Book). Refer to gynae OPC.

80. Ectopic pregnancy

81. Recurrent miscarriage

82. Urinary incontinence

83. Trichomonas vaginalis STI

84. Infertility and preconception advice

85. Severe dyskaryosis – abnormal pap smear

86. Intermenstrual bleeding

87. Polycystic ovarian disease

BMJ 29 Sept 2007 Managing anovulatory infertility and PCOS A Balen et al. Metformin (BMI > 35) is less effective than clomifene as 1st line tx in infertile PCOS. Fertility rates: metformin (7.2%) vs. clomifene (22.6%) vs. clomifene + metformin (25.5%). Laparoscopic ovarian diathermy induction of ovulation 34% vs. 67% with FSH.

88. Endometriosis
    BJGP June 2007: Mx of endometriosis in general practice Z Pugsley et al. Prev is 10%. Median delay of 9 yrs in diagnosis from repeated consultations and negative ixs. 1/3 saw GP ≥ 6 x before diagnosis. US only helpful in diagnosis in 10.6%. 39% referred to gynae ≥ 2x before +diagnosis (laparoscopy). Sxs: dysmenorrhoea, pelvic pain, deep dyspareunia, pain with defecation, during micturition and subfertility.

    BMJ 3 Feb 2007: Clinical review: Endometriosis C Farquhar. Coc, oral or depot medroxyprog and LNG-IUS as effective as GnRH analogues and can be used long-term. Lap excision or ablation at time of diagnostic lap if possible. Endometriata are best stripped out vs. drain and ablate. 5 yrs postop or med tx, 20-50% recur. Sxs: dysmenorrhoea, dyspareunia, cyclic pelvic pain, subfertility, painful defecation during menses, adnexal mass, chronic LBP/ abdo pain. Diagnosis= laparoscopy and not CA125 or TV U/S. Refer if fail rx coc + NSAIDs.

89. Contraception advice

90. Post-partum contraception

91. Pre-eclampsia

92. BMJ 7 April 2007: Screening programmes for chlamydial infection: when will we ever learn? Nichola Low. Acceptance of the effectiveness of chlamydia screening programmes in Sweden and US before balance of benefits and harms was understood.

Despite screening, rates in Sweden, Europe and US increasing. National Chlamydia screening programme (NCSP) in phase 3 2008 but may need a more proactive than opportunistic approach. Main burden of infection16-19yo F and 20-24yo M.

93. BMJ 24 Mar 2007: Imaging the endometrium in PMB. A Sahdev. RFs for endometrial CA: >5yrs of unopposed oestrogen HRT, tamoxifen, hereditary nonpolyposis colorectal CA, obesity + DM, HTN, endogenous or exogenous ↑ in oestrogens. TV U/S assess endometrial thickness ≥ 5 mm requires bx. ≤ 5 mm NPV for CA 98%. MRI preop staging CA.

94. BMJ 26 May 2007 Clinical review: Genital Herpes and its mx. P Sen et al. Ix: PCR (high sensitivity and specificity, type specific), serum DFA or Elisa Ag detection (low sensitivity, high specificity), culture (high sensitivity and specificity, type specific). Rx: 5-10 d of: acyclovir 200 mg 5x/d or 400 mg tds or valaciclovir 500 mg-1g bd or famciclovir 250 mg tds
Suppressive rx to HIV and HSV: acyclovir 400 mg bd or valaciclovir 250 mg bd or 500 mg-1g od or fam 250 mg bd. ↓HIV RNA genital shedding and plasma viral load.

Genital Herpes in a pregnant F: 1st episode in 1st or 2nd trimester – rx oral or IV acyclovir + NSVD. In last 4 weeks pregnancy, continuous acyclovir and after 34/40, C-S. If recurrent HSV+ no lesions, then NSVD and daily suppress acyclovir in last 4 weeks. If lesions, C-S. HSV2 ↑ risk 2x of HIV!

95. <u>70 yo M brought in by wife, early Alzheimer's dementia (NICE ALZHEIMER'S DEMENTIA, Nov 2006)</u>

<u>DG</u>: restless, forgetful, irritable, other sxs (HA, gait, confused, fits, depressed, personality).

<u>DDx</u>: discrete (toxic confusional, head injury, fit, EtoH, cerebral hypoxia, transient global amnesia, post ECT, PTSD, fugue state, focal retrograde amnesia) vs. persistent (drug toxicity, amnesic syndrome, semantic dementia, AD or #1 vascular dementia)

<u>RF</u>: diet, CVA, FH.

<u>MMSE</u> (address, age, DOB, year, time, place, person recognition, year of WWII, PM, count back 20 to 1.) Diagnosis ≤ 6.

<u>Addenbrooke's Cognitive Exam</u> (patient details (name, age, DOB), current day of the week, month, year, season, date. Time to the nearest hour without checking a watch. Ask them to repeat three objects after you (lemon, key, ball), then ask serial sevens or months of the year backwards, or 'world' backwards, then ask for the three objects (lemon, key, ball). Ask them to name two objects (pen, watch) and a constituent part (strap, hands, nib, lid). Write a sentence. Repeat 'no ifs, ands or buts'. Follow a written instruction (close your eyes). Follow a three part spoken instruction (pick up paper in right hand, fold in half and place on table). Draw a clock face, time 13.50hrs. Copy the goemetric diagram (overlapping pentagons or 3-d cube). Ask about current prime minister, woman PM and assassinated US president. ACE includes the seven part address, extra drawing, counting blobs in squares, pronunciation of written and spoken words, interpretation of obscured letters, comprehension test (matching pictures to themes), naming tests and 'how many words in one minute' beginning with letter P and then how many animals in one minute.

IPS: Concern re Alzheimers. IADL, family, social isolation. EdxCU. CMx: Options: ix: CRP, FBC, TFT, VDRL, B12/ folate, U/E, MSU, refer to memory clinical, MRI 62%, EEG 31%, CT 29%, gingko biloba if just forgetful, CMHT, neuro to diagnosis (neuropsychological test premorbid IQ, current IQ, memory, executive function (frontal lobe info retrieval), language and visuospatial (parietal)) + tx, psychogeriatrician, self help group. Memory aids: post-its, white board (pt puts Q and carer puts answer on it), filofax, PDAs, alarms, tape record and movement activated ('have you turned off the gas? Have you taken your keys?) LF. FU with family. Home visit. Seek carer views of baseline condition, SocS. DTA. Admit?

NICE ALZHEIMER'S DISEASE Nov 2006

Three Acetylcholinesterase inhibitors (Donepezil, galantamine, rivastigmine) for mild (MMSE 21-26), mod (10-20) to moderately severe (10-14) AD. Only psych, neurologists or care of the elderly should start tx. Review q 6/12 by MMSE score and global, functional and behavioural assessment. Continue rx while the MMSE score remains $\geq$ 10 points and global, functional and behavioural condition remains at a level where the drug has a worthwhile effect. Memantine for mod severe to severe (< 10) AD, only as part of clinical studies.

BMJ 9 Feb 2008: Assessing mental capacity: the Mental Capacity Act T Nicholson et al 5 principles of MCA:
1. Capacity should always be assumed. A person's diagnosis, behaviour or appearance should not lead you to presume capacity is absent. 2. A person's ability to make decisions must be optimised before concluding that capacity is absent, All practicable steps must be taken, i.e. giving sufficient time for assessments; repeating assessments if capacity is fluctuating; and, if relevant, using interpreters, sign language, or pictures. 3. Patients are entitled to make unwise decisions. It is not the decision but the process by which it is reached that determines if capacity is absent.

4. Decisions (and actions) made for people lacking capacity must be in their best interests. 5. Such decisions must also be the least restrictive option (s) for their basic rights and freedoms.

Advance decisions: can be verbal but if refusing life sustaining txs (i.e. ventilation) must be written, signed and witnessed to be valid. Section 5 of the MCA protects from legal liability. Court of protection (health and welfare decisions).

96. <u>NICE June 2006 Diagnosis and Mx of Parkinson's Disease in Primary Care</u>

PD common, chronic, progressive neurodegenerative movement disorder resulting from the death of the dopamine-containing cells of the substantia nigra. The diagnosis is 1° clinical, based on a hx and exam. Classically present with bradykinesia, rigidity and rest tremor. Parkinsonism can also be caused by drugs, multiple cerebral infarction and degenerative conditions, progressive supra-nuclear palsy (PSP) and multiple system atrophy (MSA).

PD frequently develop psychiatric problems, i.e. depression and dementia. Autonomic disturbances and pain (which is rarely a presenting feature of PD) may later ensue, and the condition progresses to cause significant disability and handicap with impaired quality of life for the affected pt. Family and carers may also be affected indirectly.

<u>Suspected PD should be referred quickly and untreated to a specialist with expertise</u> in the diff diagnosis of this condition. Suspected mild PD should be seen within 6/52 but new referrals in later disease with more complex problems require an appt w/n 2 weeks. The GDG advise that patients diagnosed with PD should be seen at regular intervals of 6–12 months to review their diagnosis and reconsidered if atypical clinical features develop. Acute levodopa and apomorphine challenge tests should not be used in the diff diagnosis of parkinsonian syndromes.

Regular access to specialist nursing care-clinical monitoring and medication adjustment, a continuing point of contact for support, including home visits, when appropriate a reliable source of info about clinical and social matters of concern to people with PD and their carers which may be provided by a PD nurse specialist.

Access to physio should be available for: gait re-education, improvement of balance and flexibility, enhancement of aerobic capacity, improvement of movement initiation, improvement of functional independence, including mobility and ADL, provision of advice regarding safety in the home environment.

Access to OT for: maintenance of work and family roles, employment, home care and leisure activities, improvement and maintenance of transfers and mobility, improvement of personal self-care activities, eating, drinking, washing and dressing, environmental issues to improve safety and motor function, cognitive assessment & appropriate intervention

Access to speech and language therapy for: improvement of vocal loudness and pitch range, including speech therapy programmes ie Lee Silverman Voice Treatment (LSVT), teaching strategies to optimise speech intelligibility, ensuring an effective means of communication is maintained throughout the course of AD, use of assistive technologies, review and management to support the safety and efficiency of swallowing and to minimise aspiration.

Palliative care requirements should be considered throughout all phases of the disease. Give PD and their carers the opportunity to discuss end-of-life issues with appropriate healthcare professionals.

97. BMJ 21 April 2007 Clinical review: Delirium in older people J Young et al. DMS IV: Disturb of LOC w/ ↓ ability to focus, sustain or shift attention. Changed cognition. Develops in a short period of time and fluctuates over day caused by medication, intoxication, general condition or > 1 aetiology. Confusion

assessment method: acute onset and fluctuating course, inattention, disorganised thinking, altered LOC. RF: > 65, multiple disease, dementia, infection, dehydration, EtoH, malnutrition. DDxs: LRTI, UTI, constipation, lyte disturbance (ARF, dehydration, hyponatremia, drugs ((anticholinergics – antihistamines, antispasmodic, TCA, benzo, codeine, theophylline, frusemide, antiparkinsons, digoxin, oxybutinin) or psychoactive), EtOH withdrawal, pain, neurological (CVA, fit, subdural haematoma, hypoxia, sleep deprivation, NOF operation).

98. BMJ 14 Oct 2006: Investigating iron status in microcytic anaemia M Galloway et al ≥ 20% of elderly have MCV < 75 and not Fe deficiency. Ferritin < 15 confirms, > 100 r/o. Trial of Fe rx if ferritin ≤ 40. Ix? Gastroscopy, colonoscopy.

99. Benign Prostatic Hypertrophy

BMJ 9 September 2006 BPH: tx in 1°care AK Patel et al α-antagonist improve LUT sxs w/n 48h. Alfuzosin and tamsulosin od are safest in aged (s/e postural hypotension, dizzy, HA). 5-α-reductase inhibitors (dutasteride, finasteride) ↓prostatic volume by 20-30% but take 6/12 to improve sxs and are more effective in larger prostates who are at high risk of disease progression (s/e erectile dysfunction, gynaecomastia, ↓libido, ↓PSA concentration by 50% so adjust if suspect CA). Long-term combo (α-antagonist+5α reductase ↓disease progression in high risk). NICE: refer suspected complications (recurrent UTI, haematuria, retention, renal impair, hydronephrosis, CA?, large residual volumes (> 200 ml), no improvement on initial medical treatment.

100. Requests PSA (common GP request)

DG: sxs? Understanding PSA? IPSS (int'l prostate sys score) - incomplete emptying, frequency, intermittency, urgency, weak stream, straining, nocturia? FH prostate CA? OE bladder, PR prostate, chaperone. IPS: ICE. EdiagnosisCU: BPH. PSA test. CMx: Options: wait see, test PSA (refer if > 4, suspicious prostate on exam, haematuria, obstructive uropathy, residual urine >300 ml, recurrent UTI, bladder stone, impaired renal function. PSA 4-10 70% FP.) Rx? Ix - MSU, PSA, U/S, U/E, uroflowmetry, TRUS+ biopsy to r/o CA. FU. DTA. PILS.

BJGP April 2007 It's a maybe test: men's experiences of PSA testing in 1° care. R Evans et al. Decision for test affected by social + media, not pt-led. Created uncertainty even if normal PSA. ↑PSA led to further ixs + anxiety.

BJGP Oct 2006: Clinical features of prostate CA before diagnosis: a pop-based case-control study. W Hamilton et al. 21 surgeries in Devon. 217 prostate CA patients '98-'02. PPV: urinary retention, impotence, frequency, hesitancy, nocturia, haematuria, weight loss, abnormal rectal exam deemed benign and deemed malignant.

BMJ 25 Nov 2006 Clinical review: Clinically localised prostate CA T Wilt et al. 80% cases > 65 yo. Cut-off value 4% PSA for normal but lower levels also in CA. Low risk PSA ≤10, Gleason score ≤6 and clinical stage T1c or T2a. Intermittent risk PSA >10-20, Gleason 7, or clinical stage T2b. High risk PSA >20. Gleason 8-10 or clinical stage T2c.
Tx options: watchful waiting (repeated DRE, PSA, prostate biopsy) survival 10 y; radical prostatectomy (long-term urinary incontinence, urethral stricture, bowel and erectile dysfunction); EBRT external beam radiation tx may not eradicate CA (incontinence, tx related death, diarrhoea, erectile dysfunction, urethral stricture, bleed); brachytherapy radioactive implants (urinary retention, incontinence, impotence, urethritis); androgen deprivation tx (gynaecomastia, impotence, osteoporosis, lost libido, hot flush); cryoablation (impotence, pelvic pain). M with life expectancy of 10-15 yrs unlikely to benefit from testing.

Doctor Mag 26 Sept 2006 PSA tests: harm vs. benefit C Parker Royal Marsden.≥50% of all screen-detected prostate CA would never have caused any sxs, even without tx. Early detection vs. anxiety and unnecessary txs (risk impotence and incontinence) for irrelevant prostate CA. Suggest active surveillance vs. immediate radical tx. Cochrane review no evidence for/ against PSA screening. Selective screen ↑risk: +fhx prostate CA, black, abnormal prostate on DRE. Recheck ↑PSA several weeks later as temporal variation. Prostate Cancer Prevention Trial 5000 M showed no good evidence as to what constitutes ↑ PSA.

101. Pre-test HIV counselling – 30yo M

DG: why? Gay. IVDA IPS: Consent. Confidential. Edx-CU: Ab to HIV vs. AIDS. 99.7% sensitive. 3/12 window. CMx: Options: Private: p24 Ag blood test at 4/52. Test GUM or here. National

Aids Helpline. Terence Higgins Trust helpline/website. Implications: insurance, mortgage, partner. PILS. FU.

102. Demands Viagra rx

<u>DG</u>: organic vs. psycho impotence? Morning erection? Stamp test? Endocrine d/o - NHS rx. Drugs - b-blocker, thiazide, nitrates, spironolactone, phenytoin, TCA, recreational drugs. Fhx DM? OE: examine testes r/o 1° hypogonadism. <u>IPS</u>: Taxi driver. New young F partner. Impact on relationship. Explain diagnosis. Check understanding. <u>CMx</u>: Options: Ix - FBC, lipids, lft, u/e, FBG, TFT (if sx), TT (normal total testosterone 12-29 nmol/l, measure between 8 and 11 am. If low do: SBHG, PRL, LH + FT (normal free 250-550 pmol/l)); viagra private rx - 50 mg weekly (#4), refer for psycho-sexual therapy, refer to urologist (penile implant, caverject, MUSE (transurethral alprostadil - medicated urethral system of injection), rx for 1° hypogonadism (atrophic testes) testosterone undecanoate 1,000 md q 8/52 + tadalafil 20 mg after 3/12 on testosterone. PDE5 inhibitors only work with normal testosterone levels. ED in 40s refer to cardiology for CT angiography (CAD). HTN in ED rx ARB (valsartan) instead of ACEI. 75% of T2D have ED so measure TT. More visceral fat so need more testosterone to $\uparrow$ activity of lipoprotein lipase. $\downarrow$ testosterone by 1%/yr after 30 yo. The BLT-T2D study: low TT associated with BMI, waist circumference and HbA1C. ED 77% inversely associated with TT and FT. Testosterone supplementation in T2D and metabolic synd improves insulin resistance, reduces visceral fat mass and HbA1C.

EYES/ ENT/ DERMATOLOGY

103. Fundoscopic exam – proliferative diabetic retinopathy

104. Fundoscopic exam – headache, pain behind left eye –
     glaucoma

105. Fundoscopic exam – hypertensive retinopathy

106. Fundoscopic exam – papilloedema

107. BMJ 10 Feb 2007: Txs for neovascular acute macular
     degeneration. U Chakravarthy et al. VISION study. Intravitreal
     injection of pegaptanib sodium (selective antagonist of vascular
     endothelial growth factor) to prevent choroidal neovascularisation.

108. BMJ 15 July 2006 Cataract and surgery for cataract D Allen.
     Commonest operation in developed world and #1   cause   of
     blindness in developed world. RF: tobacco, DM, systemic steroids,
     malnutrition, UVB.

109. Sudden diminished hearing

     DG: denies right ear pain. IPS: ICE? Job? Primary school teacher.
     OE: TMs nad. Tuning fork = unilateral SNHL, interpret pure tone
     audiogram. EdxCU: mumps vs. acoustic neuroma. CMx: Send in
     for acute admission to ENT: MRI internal acoustic meatus and
     carbogen tx (administer within 24h to save hearing), mumps titres.

110. 30 yo pregnant F presents with bilateral hearing loss (Classic
     textbook presentation of otosclerosis DG: pregnant, hears better in
     noisy environment, fhx of hearing problems. IPS: ICE. OE. Tuning
     fork – conductive hearing loss, normal TMs, blue sclerae. EdxCU:
     otosclerosis. CMx: refer for stapedectomy. LF.

111. Otitis externa DG: uses cotton buds. Fhx of deafness? Swim? Had grommets as child. IPS: ICE. OE: narrow EAM, white D/C. EdxCU. CMx: Swab? Pope wick + sofradex drops or otomize spray if fits. Dirty water: consider ciprofloxacin for pseudomonas cover but warn of rare photosensitivity rash.

112. BMJ 2 June 2007 Sinusitis N Chadra et al. > 7d, purulent, F, facial pain, complications (periorbital cellulitis, orbital abscess, meningitis, osteomyelitis, cerebral abscess. Rx steam inhalation, analgaesia, pseudophedrine, xylometaz, intranasal CS, 2/52 amoxicillin (trimethoprim or cefaclor if allergic) or co-amox (clarithromycin if allergic). Refer if > 12/52, I/C for CT.

BMJ 17 Feb 2007 Clinical review: Sinusitis and its mx. K Ah See et al. Chronic rhinosinusitis refer ENT for FESS. Urgently refer complications: orbital sepsis or intracranial sepsis. Common causes: viral, allergic, anatomy, tobacco, DM, swim, dive, dental infections/ procedures. Rare causes: cystic fibrosis, CA, mechanical ventilation, NG tubes, Samter's triad (aspirin sensitive, rhinitis, asthma), sarcoidosis, Wegener's granulomatosis, immune deficiency, sinus op, immotile cilia syndrome.

Acute < 4 weeks, subacute, chronic> 12 weeks. Rx antibiotics. Little evidence intranasal steroids, decongestants. Chronic rx: topical nasal steroids and antibiotics to cover Gram negative and anaerobes.

Refer if fail 3/12 med tx. Complications: Pott's puffy tumour (OM of frontal bone), optic N compression, cavernous sinus thrombosis through haematogenous spread through superior ophthalmic veins or pterygoid venous plexus.

113. Meniere's disease: DG: 50 yo F c/o dizzy, N/V and hearing loss. Refer to ENT for PTA to confirm diagnosis. CMx: refer for vestibular rehabilitation and may rx vestibular sedatives prn.

114. BJGP Jan 2007 Use of antibiotics for sore throat and incidence of quinsy N Dunn et al. Case control retrospective GP research database 95-97. 606 recorded cases of quinsy but only 192 presented as uncomplicated sore throat. RFs: smokers, 21-40 yo and M. Median 2d after tonsillitis and 3 days after sore throat. Trend for antibiotics for tonsillitis cases but no evidence that antibiotics prevent quinsy for those labelled sore throat or pharyngitis. 2/3 quinsy case did not see GP for sore throat. Low dose pen 250 mg qds rx in GP.

115. BMJ 23 Sept 2006 Clinical review: Halitosis SR Porter et al. Poor oral hygiene #1 gram negative bacteria. In absence of objective (detect sulphides, gas chromatography) oral malodour psych halitophobia. Systemic causes: F, URTI, H pylori, GERD, fetor hepaticus, end stage renal disease, DKA, leukaemia, menstrual breath. Tx: oral hygiene, tongue clean, mouthwash (chlorhexidine).

116. Give advice to mother on treatment of acne
BMJ 4 Nov 2006 Clinical review: Acne S Purdy et al. Treatment assess after 6/52 and if beneficial, continue for 4-6/12. Start with topical benzoyl peroxide for mild, then topical clinda or erythro. RCT topical erythro+benzoyl as effective as oral oxytetracycline and minocycline for mild acne. Mod acne that fails to respond to topical, try oral antibiotics (od tetracycline (not minocycline – irreversible pigmentation and ↑ liver damage) supplement with topical benzoyl peroxide. Try alternative if no response after 6/52. Consider yasmin/ dianette. Mod-severe: refer for isotretinoin, laser tx or phototx.

117. BMJ 9 June 2007 Clinical review: Herpes zoster D Wareham et al. RF for postherpetic neuralgia after HZ (> 50, F, prodrome, severe/ disseminated rash, severe pain, PCR detectable VZ virus viraemia). Complications of HZ ophthalmicus (conjunctivitis, episcleritis, scleritis, keratitis, iridocyclitis, choroiditis, papillitis, III palsy, retinitis, optic atrophy). Rx for IC adults: acyclovir (800

mg 5x a day for 7d w/n 72h rash), famciclovir (750 mg od), valaciclovir (1g tds x 7d to ↓acute pain and development of PNH), brivudin (Europe), prednisolone (↓acute pain, ↓from 60 to 30 mg od after 7d, then 15 mg for 7d), amitriptylline (↓incidence of PNH, ECG pretx)

118. BMJ 19 August 2006 Clinical review: Psoriasis and its mx C Smith et al. Gene locus PSORS-1 on chromosome 6p2 is major determinant of psoriasis. 2-3x↑CVD and lymphoma and nonmelanoma skin CA after excessive photochemotherapy. 25% anxiety/dep. Topical tx: corticosteroids potent and rapid onset. Can use combined with NSAID (calcipotriol) to maintain remission and risk from continuous use (cutaneous atrophy and rebound or pustular psoriasis. Vitamin D analogues as potent as steroids but skin irritant and slow onset. Narrow band UVB and PUVA (skin CA risk) short-term treatment. Methotrexate ideal for mod-severe (arthritis, nail) needs routine liver biopsy to r/o silent liver fibrosis and cirrhosis. Serum procollagen III q 3/12 during treatment as a surrogate marker for liver toxicity to avoid liver bxs. ? cytokine TNFα targeting drugs (etanercept, infliximab and agents that target T cells or Ag presenting cells (efalizumab). Refer if fail, 3/12 topical.

119. Atopic Eczema (NICE May 2007)
Adopt a holistic approach when assessing a child's atopic eczema, taking into account the severity of the atopic eczema and the quality of life (everyday activities and sleep) and psychosocial wellbeing. There is not necessarily a direct relationship between severity of the atopic eczema and impact of the atopic eczema on quality of life.

Skin/physical severity		Impact on quality of life and psychosocial wellbeing	
Clear	Normal skin, no evidence of active atopic eczema	None	No impact on quality of life
Mild	Areas of dry skin, infrequent itching (with or without small areas of redness)	Mild	Little impact on everyday activities, sleep and psychosocial wellbeing
Mod	Areas of dry skin, frequent itching, redness (with or without excoriation and localised skin thickening)	Mod	Moderate impact on everyday activities and psychosocial wellbeing, frequently disturbed sleep
Severe	Widespread areas of dry skin, incessant itching, redness (with or without excoriation, extensive skin thickening, bleeding, oozing, cracking and alteration of pigmentation)	Severe	Severe limitation of everyday activities and psychosocial functioning, nightly loss of sleep

When clinically assessing children with atopic eczema, identify potential triggers: irritants, soaps and detergents (shampoos, bubble baths, shower gels and washing-up liquids); skin infections; contact allergens; food allergens; inhalant allergens. Consider a diagnosis of food allergy in children with atopic eczema who have reacted previously to a food with immediate sxs, or in infants and young children with moderate or severe atopic eczema that has not been controlled by optimum mx, particularly if associated with gut dysmotility (colic, vomiting, altered bowel habit) or FTT. Tailor the tx step to the severity of the atopic eczema. Emollients should form the basis of atopic eczema mx and should always be used, even when the atopic eczema is clear. Mx can then be stepped up or down, according to the severity of sxs, with the addition of the other txs (see below).

Offer info on how to recognize and mx flares of atopic eczema (↑ dryness, itching, redness, swelling and general irritability), stepped-care plan, and prescribe txs that allow children and parents to follow this plan. Offer a choice of unperfumed emollients to use every day for moisturising, washing and bathing. May include a combo of products or 1 product for all purposes. Leave-on emollients should be prescribed in large quantities (250–500 g weekly) and easily available to use at nursery, pre-school or school.

Mild atopic eczema	Moderate atopic eczema	Severe atopic eczema
Emollients	Emollients	Emollients
Mild potency topical corticosteroids	Moderate potency topical corticosteroids	Potent topical corticosteroids
	Topical calcineurin inhibitors	Topical calcineurin inhibitors (tacrolimus and pimecrolimus) 2nd line > 2 yo
	Bandages	Bandages (occlusive wet wrap)
		Phototherapy
		Systemic tx (antihistamines) 1/12 trial

Use mild potency for the face and neck, except for short-term (3–5 days) use of mod potency for severe flares. Use moderate or potent preps for short periods only (7–14 days) for flares in vulnerable sites (axillae and groin). Very potent preps need specialist dermatological advice.

Offer info on how to recognise the sxs and signs of bacterial infection with staphylococcus and/or streptococcus (weeping, pustules, crusts, atopic eczema failing to respond to therapy, rapidly worsening atopic eczema, fever and malaise). Provide clear info on how to access appropriate tx.

Offered info on how to recognize eczema herpeticum. Signs: areas of rapidly worsening, painful eczema; clustered blisters consistent with early-stage cold sores; punched-out erosions (circular, depressed, ulcerated lesions) 1-3 mm uniform (may coalesce to form larger areas of erosion with crusting); possible fever, lethargy or distress. Educate children with atopic eczema and their parents or carers about atopic eczema and its tx.

Referral for specialist dermatologist if: The diagnosis is, or has become, uncertain. Mx has not controlled the atopic eczema satisfactorily based on a subjective assessment by the child, parent or carer (the child is having 1–2 weeks of flares per month or is reacting adversely to many emollients). Atopic eczema on the face has not responded to appropriate treatment. The child or parent/carer may benefit from specialist advice on treatment application (i.e., bandaging techniques). Suspect contact allergic dermatitis (persistent atopic eczema or facial, eyelid, hand atopic eczema). The atopic eczema is causing significant social or psychological problems for the child or parent/carer (sleep disturbance, poor school attendance). Atopic eczema is associated with severe and recurrent infections (deep abscesses or pneumonia).

RESPIRATORY SYSTEM

120. Pancoast tumour

DG: 65 yo M weight loss and persistent cough. OE: ptosis, miosis, anhydrosis (Horner's). Interpret cxr: mass in left apex with rib destruction and consolidation. EdxCU: Pancoast tumour = lung CA.BBN. Refer urgently for radiotherapy and surgery. Most are non small cell CA (SqCCA).

121.Difficulty breath 40 yo F

DG: off long haul flight from Los Angeles. Chest pain. Acute SOB. Tall slender. Smoker. IPS: ICE. OE: no breath sounds on right, tachycardic, diaphoretic, deviated trachea to left. EdxCU: Tension pneumothorax. CMx: blue light ambulance. Administer 100% oxygen. Needs urgent needle (14G-16G) decompression in 2nd ICS. CXR to confirm re-expansion (not to make diagnosis).

122. Wheeze in 15 yo F - new-onset asthma

DG: wheezy, nocturnal cough, SOB, frequency, sports, dust mites, pets, tobacco. FH asthma. OE lungs, PFM.
IPS: ICE, stigma EdxCU: asthma, inhalers, peak flow meter. RGO.
CMx: Options: Rx salbutamol (blue reliever) +/- beclomethasone (brown preventer) inhaler + home PFM, chart PFR bd. Acute attack mx. PILS.

BJGP Mar 2007 House dust mite allergen avoidance and self-mx in allergic patients with asthma RCT MP de Vries et al. 126 patients house dust mite impermeable vs. placebo bed covers. No ↓ in use of inhaled steroids, pfr, dysnoea, wheeze or cough.

Doctor Mag 24 Oct 2006 Wheezy infant S Wilson, paeds fellow. Viral-induced wheeze < 3 yo, outgrow by mid childhood. Fhx of atopy? Asthma? CF if clubbing, suboptimal growth, GI sxs, persist wet cough? Inhaled salbutamol 400-600 mcg thru spacer or if needs > 3-4x a day, add oral prednisolone (1mg/kg/day). If < q6-8 weeks, then no inhaled steroids. If nocturnal wheeze or uses salbutamol > 3x a week, add budesonide or beclomethasone 200-400 mcg/d through spacer. If doubt wheeze? FB, trial 600 mcg spacer. If atopy, skin prick, fbc, IgE. CXR, sweat test?

123. Pneumonia – fever, cough, SOB

124. Lung CA BJGP Aug 2006 Negative CXRs in 1° care patients with lung CA S Stapley et al. 23% had negative x-rays w lung CA (38 of 164). 38,000 UK patients are diagnosis with lung CA/year. Recommend assess sx change.

125. Administration of O_2 therapy for respiratory distress – COAD

BMJ 1 July 2006 ABC of COPD. O_2 and inhalers. G Currie. Exacerbation - venturi FM 24% or 28%. LTOT PaO_2 < 7.3 kPa on two separate occasions at least 3 weeks apart during clinical stability OR PaO_2 7.3-8 kPa and evidence of 2^0 polycythaemia, pulmonary HTN, peripheral oedema or nocturnal hypoxaemia. Use for ≥ 15h/day. Short burst O_2 tx controversial. Cabin pressure ↓ O_2 for patients with COPD. In flight O_2 if < 92% on pulse oximetry.

BMJ 28 Jan 06 Oxygen tx at home G Gibson. From Feb 1 06, 3 changes: all forms of home O_2 tx will be provided by a single supplier in each region of UK after receipt of a home O_2 order form specifying details of usage, flow rate (2 lpm till assessed), and expected hours of use; ambulatory O_2 (including liquid) and hospital specialists may order home O_2 directly. 63000 patients on cylinders. Indications are unchanged.

126. COAD

BMJ 14 April 2007 COPDA McIvor et al. barrel chest, height, weight, spirometry (gold standard), stop smoking, annual flu, pneumococcal jab q5-10 yrs, bronchodilators, short acting β-agonists prn, LABA, ICS or as combo with LABA.

BMJ 15 July 2006: ABC of COPD. Ventilatory Support. Non invasive ventilation (NIV) Insp+ airways P up to15-20 cm H2O helps offload tiring muscles and ↑ elimination of CO_2 and expir PAP at 4-6 cm H2O helps splint airway and flushes CO_2 from mask. Overcomes intrinsic + end exp P thus ↓ atelectasis and ↑end TV. ↓ mortality, need for intubation, tx failure and complication rate. Used in hosp wards and start in A+E.

BMJ 8 July 2006 ABC of COPD Acute Exacerbations G Currie.
10% of all medical admissions to UK hospitals. Exacerbated by bacteria, virus or pollutants. Type 2 (↓ O_2, ↑CO_2, ↑ or normal HCO_3, pH normal or ↓) give 24% or 28% controlled O_2 via Venturi FM. Type 1 respiratory failure (↓ O_2, normal or ↓ CO_2, normal HCO_3, pH normal or ↑) = titrate O_2 up to maintain > 90% sat. Permits detection of ↑ CO_2 concentration or ↓ pH due to loss of hypoxic drive. Bronchodilators (inhaled salbutamol). Oral steroids. Aminophylline controversial. NIV for hypercapnic respiratory failure (pH 7.25-7.35). Low MW hep.

127. Difficulty breathing - 2° mets, pleural effusion

128. Pulmonary Tb

BJGP Feb 2007 Unwrapping the diagnosis of Tb in 1° Care: a qualitative study. E Metcalf et al. Interviewed 17 patients +16 GPs. RF: I/C (HIV), chronic EtOH, drug misuse, young/old age, minority ethnic/ immigrants/ travel in ↑ incidence areas. Traditional sxs: persist cough, haemoptysis, night sweats/ F, weight loss, CW pain, non-resolving pneumonia, dyspnoea. Ixs: cxr, blood, sputum. Delays: unclear diagnosis, lack of continuity of care + patient/doctor suboptimal communications.

BMJ 20 May 06 Pulmonary TB: diagnosis and tx. I Campbell et al. RF immigrants, old, IC patients, poor living conditions. In developing countries check sputum for AFB if cough > 3/52 despite broad spectrum antibiotics.

CXR unspecific for IC patients. +tuberculin test (absent previous infection or BCG vaccine) ↑ Tb probability even if sputum is negative. Standard tx (6/12 rifampicin and isoniazid), with initial 2/12 of pyrazinamide + ethambutol. Cross infection if patient is sputum + for AFB on a direct smear. Prompt contact tracing. Treating at home no more likely to lead to cross infection than treatment in hospital. BCG vaccine to all at high risk of tb exposure.

RENAL/UROLOGY

129. <u>BMJ 14 Oct 2006: Clinical review: Acute renal failure</u> R Hilton
#1 cause in-hospital is acute tubular necrosis from multiple
nephrotoxic insults (sepsis, ↓BP, nephrotoxic rx, radiocontrast
media). No rx can limit progression. ARF creatinine > 50.

<u>Pre-</u> (40-70%, hypovolaemia, renal hypoperfusion (NSAID, ACEI,
AAA, RAS, hepatorenal syndrome), ↓BP (shock), oedematous
states (heart failure, cirrhosis, nephrotic syndrome)
<u>Intrinsic</u> (10-50%, glomerular, interstitial nephritis, tubular,
vascular)
<u>Post or obstructive</u> (10%, stone, papillary necrosis, urethral
stricture, BPH, CA, radiation, fibrosis, pelvic CA, retroperitoneal
fibrosis).

Assess: hx, exam, creatinine, small kidney u/s, palpable bladder,
Urinary Na, UA (red cell casts, proteinuria), fluid challenge,
macroscopic haematuria, loin pain, atherosclerotic vascular
disease, serum Igs (bence jones proteinuria = MM), CRP, CK
(rhabdomyolysis).

Tx: loop diuretic but ototoxic in high doses, DA (risk tachycardia,
peripheral gangrene), natriuretic peptides (↓BP), continuous vs.
intermittent haemodialysis.

130. <u>Haematuria – adult polycystic renal disease vs. renal calculi</u>

DG: 20 yo F 3/7 flank pain and macroscopic haematuria. UTIs?
Coagulopathies? Stones? FH? OE large NT irregular ballotable
kidney. BP 150/100. Urea 12, creatinine 160. EdxCU: adult
polycystic renal disease (autosomal dominant, counselling for
relatives of affected individuals). Ix: AXR to r/o radio-opaque
stones. Definitive ix: renal scan.

BMJ 3 Mar 2007 Clinical review: Mx of kidney stones. N Miller et al. Unenhanced helical CT is best radio ix for diagnosis urolithiasis. Shock wave lithotripsy (80-85% are simple < 2 cms calculi, proximal ureteric ≤ 1cm), ureteroscopy (pregnant, morbidly obese or patients with coagulopathies, distal ureteric stones), and percutaneous nephrolithotomy (complex renal calculi > 2cms, staghorn, stones resistant to fragmentation, proximal uret > 1cm) have replaced open op for tx. Most ureteric stones < 5mm in diameter will pass spontaneously within 4 weeks. Hx: 1°hyperparathyroidism, renal tubular acidosis, cystinuria, gout, DM, IBD, renal insufficiency, sarcoidosis, medullary sponge kidney. Horseshoe kidney, previous urinary diversion, UPJ obstruction, solitary kidney, previous renal or ureteric op, h/o pyelo, fhx of stones, drugs (triamterine, ephedrine, guaifenesin, calcium with vitamin D, indinavir or sulfadiazine, topirimate). Acute intervention: presence of infection with UT obstruction, urosepsis, intractable pain or vomiting, impending ARF, obstruction in solitary or transplanted kidney, bilateral obstructive stones. Analysis of stone: Two 24h urine collections for vol, pH, ca, oxalate, citrate, uric acid, PO4, Na, K, Mg, NH4, Cl, sulphate, creatinine; cystine screen. Ix: serum Ca, HCO3, creatinine, cl, K, Mg, PO4, uric acid, urea. In hypercalcaemia, PTH and 1,25 dihydroxyvitamin D. In cystinuric patients, as above and 24h measurement of cystine.

131. 55 yo M for blood test results – eGFR 60, GGT 150, TC:HDL 6. FHx. Diet. + EtOH 2 bottles a day. 30 cigs a day. Desk job fire marshall. Stress? Home? NZ pregnant girlfriend denied visa twice to UK, lonely. OE: BP 160/100, no murmurs. EdxCU: Chronic kidney disease stage 2. Hypertension. BNF 10 yr % = 30%. Mx: DTA. EtOH counsellor, renal scan, bone +liver profile, amylase, ECG. LF. FU.

BMJ 16 July 2007: CKD PK Mitra et al. 5-10% pop. HTN-silent CKD. CV sxs? Systemic diseases (lupus, vasculitis or MM)? NSAIDs? FHx DM, CV, HTN, PVD, and polycystic kidney. Serial measurements eGFRs, BP, weight, UA, abdo for enlarged kidney or bladder. Mx: Stage 1 or 2 and 3 with stable function: monitor renal function annually. In progressive stage 3 monitor 6 monthly. Stage 3: check hb, K, Ca, PO4, PTH and renal U/S if has lower UTI sxs, refractory HTN or unexplained progressive fall in eGFR.

Stage 4/5: refer. If proteinuria, check urine protein:creatinine ratio and refer if > 100mg/mmol. Start ACEI or ARB at 140/90. Aim for BP < 130/80. Statin and ASA if CV risk > 20%.

BJGP Dec 2006: Chronic kidney disease: a new priority for 1° care G Gomez et al. Causes of CKD: DM, HTN, acquired obstructive uropathy (prostate disease), glomerular disease (GN), adult polycystic kidney disease, reflux nephropathy.

Mx of CKD in QOF (aim BP 140/85) and NICE pending. eGFR< 30 ask 2° opinion.

UK Guidelines for Identification, Mx and Referral for CHRONIC KIDNEY DISEASE in adults

(Sources: RCGP Introducing eGFR, promoting good CKD mx; www.renal.org/ CKD guide)

The KDOQI stages of chronic kidney disease are:

Stage	GFR	Description	Treatment stage
1	90+ml/min/ 1.73 m2	Normal kidney function but urine or other abnormalities (i.e. known to have proteinuria, haematuria (no urological cause), microalbuminuria (DM), polycystic	Observation, control of BP, eGFR urine PCR if dipstick protein present; yearly

		disease or reflux nephropathy)	
2	60-89	Mildly reduced kidney function, urine or other abnormalities point to kidney disease	BP control, monitor as per stage 1. No further testing if eGFR alone.
3	30-59	Moderately reduced kidney function	also Hb, K, phosphate, Ca 6 monthly (12 if stable < 2 mL /min change eGFR over 6/12)
4	15-29	Severely reduced kidney function	Also bicarb, PTH. Plan for RRR. 3 monthly (6 if stable CKD stage 4) Refer urgently.
5	14 or less	Very severe, or endstage kidney failure (sometimes call established renal failure)	As per stage 4. Refer urgently.

eGFR 100% kidney function = 100 ml/min based on creatinine, gender, age (falls with age). Multiply by 1.2 for Afro-Carribean. Does not apply in pregnancy, acute renal failure nor to children < 18 yo.

Under-estimates severity of renal failure in people with muscle wasting (malnourished) or an amputation.

If eGFR < 60 ml/min: review all previous creatinine/ eGFR results to assess rate of deterioration. Review meds, (NSAIDs, antibiotics, mesalazine, diuretics, ACEIs/ ARBs. Test urine for haematuria and proteinuria. If protein present, request urine protein/ creatinine ratio. Assess clinically: for urinary sxs, palpable bladder, BP, sepsis, heart failure, hypovolaemia. Repeat serum creatinine within 5 days to r/o rapid progression if new finding.

Referral Criteria:
Stage 1/2: Urgent: malignant hypertension; K>7 mmol/l; nephrotic syndrome. Routine: isolated proteinuria (prot:creat ratio PCR > 100 mg/mmol); proteinuria and microscopic haematuria (PCR > 45 mg/mmol); diabetes with proteinuria (PCR> 100 mg/mmol) but no retinopathy; macro haematuria (after negative urological evaluation); recurrent pulmonary oedema with normal LV function; fall of eGFR of > 15% during 1st 2/12 on ACEI/ ARB.
Stage 3 As above plus: progressive fall in GFR; microscopic haematuria (after negative urological tests if > 50 yo); proteinuria (urine PCR > 45 mg/mmol); anaemia (after exclusion of other causes); persistently abnormal K, Ca, PO4 (unstuffed sample); suspected SLE, vacuities, myeloma; uncontrolled hypertension BP > 150/90 on 3 drugs.

Stage 4/5 Urgent. All patients should be referred and offered the options of renal replacement therapy (RRT) or conservative tx, even if RRT will not be appropriate. Exceptions may include if the CKD is part of terminal illness or function is stable and relevant tests completed and appropriate mx implemented with agreed tx plan.

Info needed on referral: general medical hx, urinary sxs, rxs, exam (BP, oedema, bladder), urine dipstick for blood and protein, urine for PCR if proteinuria present, FBC, creatinine, urea, Na, K alb, Ca, PO4, cholesterol, HbA1C (DM), all previous creatinine results with dates, results of renal scan if available.

Mx all stages: Regular clinical and lab assessment. Advice on smoking, weight, exercise, salt and EtOH intake.

CV prophylaxis: if risk > 20% at 10 yrs, consider ASA if BP < 150/90 and lipid lowering rxs (or entry into trials). Meticulous BP control. Threshold 140/90, target 130/80 in most patients; threshold 130/80, target 125/75 if urine PCR > 100 mg/mmol. Include ACEI or ARB if urine PCR > 100 mg/mmol or if diabetes and microalbuminuria present. Check creatinine and K before starting and 2 weeks after start and after each dose change. If creatine ↑ by > 20% or GFR ↓ by > 15% repeat with K and seek advice (? Stop test for RAS renal artery stenosis). If K > 6 mmol/L, check no haemolysis and check diet; stop NSAIDs and LoSalt (K containing salt substitute); stop K retaining diuretics; stop ACEI/ARB if ↑K persists.

CKD Stage 3: additional mx to include: If Hb < 11 and other causes excluded, refer for IV iron +/- ESA (erythropoietin stimulating agent) with target Hb 11-12 g/dl. Renal scan if lower urinary tract sxs, refractory hypertension, unexpected falling eGFR. Immunise against influenza and pneumococcus. Review all drugs ensure correct dose; avoid nephrotoxic drugs (NSAIDs).Check PTH level when stage 3 1st diagnosed. If high, check 25-hydroxy vitamin D and if low, give ergo- or chole-calciferol with Ca supplement (not PO4), repeat PTH after 3/12 and refer if still high.

CKD Stage 4/5: additional mx in conjunction with 2° care. Assess diet. Hepatitis B immunise. Mx of hyperparathyroidism. Correct acidosis. Info about tx. Timely dialysis access procedure. Referral/discuss even if dialysis may not be appropriate.

TELEPHONE CONSULTATIONS/PAEDIATRICS

132. <u>Mum calls: fear of meningitis in a 2 yo child</u>

<u>NICE FEVER IN CHILDREN</u>

DG: red flags: fever, non-blanching rash, neck pain, photophobia, responsive? IPS: ICE meningitis. CMx: call ambulance to pick up child or do home visit. Take down name and number.

<u>BMJ 30 Sept 2006 Clinical review: Meningococcal disease and its mx in children</u> CA Hart et al. Presents as septicaemia, meningitis or combo. Classic septicaemia: nonblanching rash (petechial→purpuric→ vascular) and F. 15% sequelae (deaf). Amputations and abnormal bone growth recently appeared as sequelae. 50% < 4 yo. Meningitis (F, HA, anorexia, vomit, drowsy, photophobia, stiff neck, + Kernig's sign, F, non-blanching rash). Mx: cefotaxime/ pen IV w/n 30 mins, Glasgow meningococcal septicaemia prognostic score ≥ 8/15 ↑risk death, IVI (20 ml/kg, up to 40 ml/kg PICU) and intropes (dopamine, dobutamine, adrenaline) for shock, paeds ICU, culture, PCR, disease notification and prophylaxis, FU hearing test. LP not in acute phase of septicaemia but 48h after physiology stable. Ix: fbc, coagulation screen, blood cult, CRP, PCR, serology, throat swab.

<u>BMJ 3 June 2006: Effectiveness of antibiotics given before admission in reducing mortality from meningococcal disease:</u> systematic review. S Hahne et al. Cannot conclude whether this effects case fatality. Severity confounding factor.

<u>BMJ 3 June 2006: Parenteral penicillin for children with meningococcal disease before hosp admission: case-control study.</u> A Harnden et al. Children given parenteral pen by GP had more severe disease on reaching hosp. Confounding factor - more severely ill are being given pen before admission so associated with a poor outcome anyway.

<u>NICE: Fever in Children – May 2007</u> <u>Meningococcal disease:</u> Non-blanching rash, with ≥ 1 of the following: an ill-looking,

lesions > 2 mm in diameter (purpura), capillary refill time of ≥ 3 secs, neck stiffness. Meningitis: neck stiff, bulging fontanelle, ↓ LOC. Convulsive status epilepticus. Children with fever, who are shocked, unrousable or showing signs of meningococcal disease, give parenteral benzylpenicillin or 3rd-generation cefotaxime or ceftriaxone at the earliest opportunity until culture results are back. Admit to paeds and assess need for inotropes. Consider refer to PICU.

133. 7-month male baby is vomiting

DG: child cries in pain with and shortly after feeding. Recent URTI. Stools loose. Third visit for same problem in 2/52. Mother not happy, still concerned. OE: well baby with soft, distended abdo. No fever. Dx: suspected intussception (prime age and male baby). CMx: refer to paeds A+E for barium enema to r/o intussusception. (M: 6-12mos). 1% UK mortality from delayed GP referral.

134. Father calls: 2-week old baby vomiting for 1 week

DG: Seen by HV, told would get better. Father says whenever he feeds baby, milk dribbles out of corner of her mouth. Baby is hungry. Dry nappy. Scant yellow stool staining. IPS: worried as baby vomiting and getting worse, cannot feed. EdxCU: concern re pyloric stenosis. CMx: home visit (assess and send patient in for scan to confirm stenosis and paeds surgical opinion.

135. BMJ 12 May 2007 Clinical review: Chronic abdominal pain in children M Berger et al. Rome III criteria 2006 (functional dyspepsia, IBS, functional abdo pain, functional abdo pain syndrome (with somatic sxs headache, limb pain), abdominal migraine (periumbilical for ≥1h, N/V, headache, photophobia)). Little known role of sexual abuse and pain. Alarm sxs: involuntary weight loss, deceleration of linear growth, GIB, significant vomiting, severe diarrhoea, unexplained F, persist RUQ or RLQ

pain, fhx inflammatory bowel dis. Tx: CBT (beneficial evidence), famotidine, added fibre (unlikely), lactose free diet (unlikely), peppermint oil (likely beneficial), pizotifen for abdominal migraine (likely beneficial), lactobaccoacillus GG (unlikely).

136. BMJ 6 Jan 2007 Clinical review: Acute gastroenteritis in children EJ Elliott Rotavirus most common. Most with mild_moderate dehydration tx with oral or enteral rehydration using_low osmolality oral rehydration solutions. Educate carers re personal + food hygiene. Mx: normal skin form retracts immediately - mx at home. 5% mild or 6-9% moderate dehydration (≥ 2 restless, irritable, sunken eyes, thirst), slow skin fold visible at 2 secs – oral rehydration at home or NG/IVI over 4-6 hours. Severe >10% dehydration: ≥ 2 abnormal sleepy, sunken eyes, drinking poorly or not at all. Skin fold ≥ 2 secs. Check acid base, urea/lytes before IVI. If shock, IV bolus, rehydrate IV over 4 to 6 hours. Refer < 6/12 baby, diarrhoea> 2 weeks, severe abdo pain, high F, poor urine output, etc.

137. BMJ 10 Feb 2007 Clinical review: Febrile seizures L Sadleir et al. Most common seizure d/o. Recurrent in 1/3 kids, low association with epilepsy. RF for epilepsy: complex febrile seizure, neurological abnormality, fhx epilepsy. Febrile seizures prolonged in 9% and tx with buccal or intranasal midazolam. Prevalence 3-8% up to 7yo. 24% fhx febrile seizure. Brief, general tonic-clonic < 10 mins in 87%. No rationale for EEG, neuroimaging or routine bloods. LP if sxs meningitis or febrile status epilepticus. Call ambulance if > 5 mins, rx rectal diazepam (0.5 mg/kg), and buccal (0.4-0.5 mg/kg) or intranasal (0.2mg/kg) midazolam. RCTs show midazolam is superior to diazepam.

138. BMJ 2 September 2006 Are the dangers of childhood food allergy exaggerated A Colver NCL (Newcastle). Incidence of severe food allergy in children is small and not↑. Risk of death is

small. Many outgrow food allergy and clinical reactivity should be reassessed periodically. Unclear what % should be prescribed adrenaline autoinjector. Only prescribe, if diagnosis made by oral challenge. 90% IgE reaction: milk, eggs, peanuts, tree nuts and seeds, fish, shellfish, soya, wheat. Optimal asthma management. Education. www.anaphylaxis.org.uk.

139. Habit cough DG: 8 yo with day-time honking, barking cough for weeks after URTI. ICE? Stress, bullying? OE: no T, chest clear. Ix: cxr, lung functions, bloods. CMx: Refer child psychologist?

LEARNING DISABILITIES

Instead of child actors, the CSA exam asks actors to act the role of a patient with a learning disability or a young teen with poor understanding.

140. Doctor Mag 28 Nov 2006 Autism J Nicholls. ASD (autistic spectrum d/o) impaired social interaction (aloof); impaired social communication (limited content of speech, not understand V/NV language); impaired imagination (repetitive vs. playing). Refer to local child development team or psych service. Asperger synd test: seek company of others? Shares interests with others? Understands feelings of others? Difficulty in following rules, taking turns? Repeat particular words/ phrases? Repeats phrases said by others? Speaks in unusual ways, monotone tone? Use invented words that do not convey ordinary meaning? Stereotyped body movements, intense attachment to objects (wheel of toy car), ↑ or ↓ sensitivity to noise, smell, T, colour, pain?

INFECTIOUS DISEASES/ACUTELY ILL

141. GP Mag 10 Nov 2006 The diagnosis and prevention of mumps T Hashmi. ssRNA paramyxovirus notifiable. Incubation 18-21 days. Parotitis, orchitis (20% M), meningitis (5-15%), transient SNHL (4% adults), spontaneous abortion. Diagnosis virus in saliva or urine from 7d prior to parotitis to 9 days after. Blood analysed for IgM or ↑ in IgG to mumps Ag. IgM in saliva is recommended ix. U/S testicular torsion. Infective prodrome to 4 days after inflammatory phase.

142. BMJ 24 Feb 2007 Clinical review: Dog Bites. M Morgan et al. Avoid 1°closure if >6h. Prophylaxis: co-amoxiclavulanic acid. Never use erythromycin or flucloxacillin alone as Pasteurella-resistant. Within 12h of injury = pasteurella multocida. High risk: IS patients, cirrhosis/ asplenia (↑capnocytophaga) or had mastectomy (lymphoedema after radiotx, DM, RA, steroid tx ↑ Pasteurella – leads to tenosynovitis in hand bites, aggressive gram negative, assoc w/ 30% mortality in septicaemia).
Travellers: consider rabies prophylaxis. Full wound exam and debride. X-ray to r/o embedded teeth, fractures, bony damage or in scalp wounds kids. Ig and tetanus toxoid if h/o ≤2 imms. Severe infections rx: imipenen with cilastatin 500 mg qds IV and clindamycin 900 mg qds IV until cultures back. If pen allergic, cipro 400 mg bd IV + metronidazole 500 mg tds IV instead of imipenem. 10-14d for cellulitis, 3/52 tenosynovitis, 4/52 septic arthritis, 6/52 osteomyelitis. Oral when CRP < 50mg/l.

143. Syphilis STI
BMJ 20 Jan 2007 Clinical review: Syphilis P French. Coiled, motile spirochaete bacterium, obligate parasite. Facilitates HIV transmission. Re-emerging in western Europe. Need high index of suspicion and low threshold for testing. 1°syph: incubation 2-3weeks (local, single painless or multiple painful papule that ulcerates, local nontender LNs, spontaneously heals 4-5 weeks), 2° incubation 6-12 weeks (general symmetric maculopapular rash palms, soles, scalp, general LNs, mucosal ulceration – snail track ulcers, condylomata lata wartlike genital lesions, F), early latent

(asymptomatic of < 2yrs), late latent (asymptomatic ≥ 2yrs), late symptomatic (3°) CV (proximal aorta – aortic medial necrosis leading to aneurysm, aortic incompetence and heart failure, coronary ostial stenosis), neurology (spinal cord syndrome - tabes dorsalis and brain general paralysis of the insane), gummatous. Ix: EIA serum test. Rx: benzyl pen 2.4 megaunits IM x1 or procaine pen 600,000 U IM x 10d or 2nd line (doxycycline 100 mg bd x14d); late latent: benzyl pen 3 injections over 2 weeks or procaine pen 900 000 U IM od x17d or doxy 200 mg bd x 28d; neurosyphilis: procaine pen 2.4U od IM x 7 days with oral probenecid 500 mg qds or doxy 200 mg od x 28 days. WHO Guidelines for mx of STIs '03. CDC '06.

144. BJGP Feb 2006: Opportunistic and systematic screening for chlamydia: a study of consultations by young adults in general practice. C Salisbury et al. Combine opportunistic and postal pop-based approaches to achieve higher coverage. Of 12,973 patients, 60% M and 75% F aged 16-24 visited once in 1-year period. 21% did not attend through postal screening, 9% attended without invite and 11.8% missed by both methods.

145. Flu-like illness - 50 yo Afro-Caribbean F (based on real pt of mine) DG: Sxs? Flu like – vague tired, cold, muscle aches. Travel hx went to Nigeria. Completed malaria tabs and had yellow fever jab. Job? OE nonspecific signs T 37C, BP 100/70, chest clear. IPS: Family concerned as she is delirious at night and does not recognise her children. EdxCU. CMx: Options: Send in to medical team. Screen family ix - blood smear for microscopy. Notifiable. Mosquito prevention. Patient was confirmed with diagnosis of malaria.

BJGP Jan 2007: Malaria in the UK: new prevention guidelines for UK travellers. B Bannister. 1500-2000 notified cases in UK. 300 million in world and 1 million deaths annually. Pakistan - chloroquine+proguanil. W and E Africa (chloroquine resistant P faciparum) – mefloquine (C/I epileptics or depression) or atovaquone+proguanil (malarone) or doxycycline. Start doxy or malarone 2 days before departure. Malarone inhibits the initial stages of parasite multiplication in liver so only need to take for 1 week on return. Newer artemisinin-containing drugs are for tx only. ABCD: awareness of risk, bite prevention, chemoprophylaxis and diagnosis+tx without delay if suspect.

BMJ 29 July 2006: Malaria: an update on tx of adults in non-endemic countries. CJ Whitty. Admit falciparum. Non-falciparum seldom admit. Mild or mod falciparum, consider oral quinine, atovaquone-proguanil or artemether-lumefantrine. Severe falciparum = parenteral quinine or artesunate. Loading dose for quinine.

BMJ 14 June 2008:Preventing malaria in travellers DG Lalloo. 5-16 UK deaths annually. ABCD. Awareness of risk (W Africa 6%, duration, activities to be undertaken), Bite avoidance (protective clothing (dusk to dawn), DEET repellent safe in kids and pregnancy, long-life insecticide treated netting), Compliance with chemoprophylaxis (C/I to mefloquine ie epilepsy or psych) and prompt Diagnosis (malaria side or rapid diagnosis tests). Chloroquine OTC, weekly dose, S/E nausea, retinal toxicity in high doses; proguanil OTC daily, SE GI, mouth ulcers if taken with chloroquine); mefloqione (private rx, weekly, se neuropsychiatric); doxycycline private rx, daily, se oesophageal sxs, photosensitivity, vaginal yeast; malarone (atovaquone + proguanil) private rx, daily, GI sxs, rash; primaquine 2nd line, private rx, side-effect: haemolysis with G6PDD.

146. BMJ 2 Dec 2006 Clinical review: Rheumatic fever and its mx A Cilliers Group A β-haemolytic strep pharyngitis.↑ in ASO titre+WHO: 2 major (carditis, polyarthritis, SC nodules, erythema marginatum, chorea) or 1 major + 2 minor (prolonged PR, arthralgia, F, acute phase reactants (ESR, CRP) manifestation. Rx: IM pen, valve replacement. Lifetime cost effectiveness of simvastatin in a range of risk groups and age groups derived from

RCT of 20, 536 pt Heart protection study collaborative group. 69 UK hosp. 40 mg x 5 years. Cost effective in a wider population than is routinely treated now, for major vascular event (nonfatal MI, coronary disease death, CVA or revascularization procedure).

147. GP Clinical 1 Sept 2006. HIV in 1°care. Brit HIV Assoc guidelines 2005, tx for 1°HIV infection, asymptomatic HIV infection CD4 201-350, asymptomatic HIV infection CD4 < 200 and symptomatic HIV or AIDS-defining illness. Tx success with HAART is viral load < 50 copies/ ml, rising CD4 within 24 weeks.

148. BMJ 29 April 06 Clinical review Diagnosis and Tx of chronic hep C infection K Patel et al. Current optimal treatment is pegylated interferon alfa and ribavirin for 24 or 48 weeks on basis of genotype and virological response. Hepatitis C virus RNA load. Transmitted by parenteral or permucosal exposure to infected blood or fluids. 70-85% chronic from acute and 20% end stage liver disease. 1-4% hepatocellular carcinoma.

149. HPV immunisation (based on school's offer to vaccinate my 12 yo) RCGP News Jan 2008: HPV immunisation – the implications for GPs, S Jarvis National Immunisation Programme for all girls aged 12. HPV 99.7% of invasive cervical CA. HPV 16, 18 (75% of cervical CA) and HPV 6 and 11 (90% of warts). Gardasil vaccine (HPV 6, 11, 16, 18) offers 98% and Cervarix (HPV 16, 18) 90.4% protection against CIN2 or higher, IM injection x 3 over 6 months.

J Fam Plann Reprod Health Care 2008; 34 (1): 3-4 HPV Vaccines: Implentation and communication issues. M Kane. Could prevent 70-80% of cervical CA. Recent data: vaccine is safe, immunogenic and effective in > 26 year-olds.

J Fam Plann Reprod Health Care 2008; 34 (4): 207-209. HPV Vaccine: Peering through the fog. Anne Szarewski. HPV vaccination 12-13 year-olds began Sept 2008, followed by 17-18 year-olds. From Sept 2009, further catch up programme for girls up to 18. Studies underway for F > 25 yo. Vaccinating boys to prevent anal CA?

<u>Know the Notifiable diseases under the Public Health Act of 1984 and Public Health Regs 1988</u>

<u>GENETICS IN PRIMARY CARE</u> (see online RCGP curriculum)

150. <u>Down's, Turner's, Klinefelter's, APKD, NF, Huntington's, hypercholesterolaemia, cystic fibrosis, Haemophilia, sickle cell, thalassaemia, haemochromatosis, fragile X, Duchenne and Becker muscular dystrophy, familial cancer, familial adenomatous polyposis, congenital deafness.</u>
<u>DG</u>: concerned about risk of inheritance. <u>EdxCU</u>: how to define gene (memory code) and mutation (faulty gene), know which ones are autosomal dominant or recessive and how to explain each and all of the conditions listed in layman's terms (one in two chance or one in four; 50% or 25%) and draw a tree diagram.
<u>CMx</u>: refer to genetic centres. Discuss online resources (OMIM, Geneclinics). Know at what age genetic testing is offered for each.

3136926R00116

Printed in Germany
by Amazon Distribution
GmbH, Leipzig